Instructor's Manual

Laboratory Manual for Principles of General Chemistry

Fifth Edition

J. A. Beran
Texas A & M University—Kingsville

John Wiley & Sons, Inc.

New York   Chichester   Brisbane   Toronto   Singapore

ISBN 0-471-00775-7

Printed in the United States of America

10 9 8 7 6 5 4 3 2 1

Printed and bound by Malloy Lithographing, Inc.

Preface

The Instructor's Manual (IM) is designed to simplify and clarify the responsibilities of the laboratory instructor, who is responsible for conducting a safe, meaningful laboratory session, and stockroom personnel, who prepare the solutions and organize the experiment's special equipment. The IM presents each experiment in the laboratory manual so that a first-time laboratory instructor can guide students toward

- developing and utilizing good, safe laboratory techniques.

- collecting data in the laboratory and using the scientific method for the data analysis.

- appreciating chemicals, their properties, dangers, and their values, influence, and significance on our technological society.

The Lecture Outline and Teaching Hints presented for each experiment have worked effectively for new laboratory instructors (many of whom are undergraduates), especially when we hold our weekly "skull" sessions.

We advise instructors to conduct a brief lecture at the beginning of the laboratory session (but not to exceed 20 minutes, including quiz time) to review various procedural, technique, and safety aspects of the experiment.

In the IM each experiment is divided accordingly:

- An **Introduction** about the significance and/or value of the experiment from the instructor's and student's viewpoints.

- A **Work Arrangement** and **Time Requirement** are suggested for the completion of the experiment.

- A **Lecture Outline** indicating the areas that require special attention before students begin the experiment.

- **Cautions & Disposal** procedures of which the instructor must be aware are given. These procedures can be presented to students or merely "watched for" during the laboratory period.

- **Teaching Hints** include brief insights into various parts of the Experimental Procedure, interpretations to questions most often asked, and expected results.

- A list of **Required Chemicals** for the experiment details an estimated amount of each chemical required *per student* (or student group). No tolerances for waste are made for the estimates. Therefore, if a large number of aliquots are used in the experiment, a 10–20% "waste" allowance (generally, a conservative waste estimate) should be made in preparing solutions. These amounts should serve only as a rough guide to preparing the solutions.

  The preparation (and preservation, wherever necessary) of each solution and the number of the experiment in which it is used, designated in parenthesis, is outlined in Appendix A of the IM.

- A list of **Suggested Unknowns** we use in our laboratories is provided. The value of the parameter (e.g., molar mass, specific heat capacity, density, etc.) to be determined in the experiment is listed for each substance as well as the composition for the mixtures we use for unknowns in our laboratories.

- The **Special Equipment** for the experiment, not normally found in the student's assigned drawer/locker, is listed. While some of the equipment may be originally stocked among the "common" equipment in some laboratories, it is only used 1–5 times during the year. We have found it less expensive to check these items out of the stockroom as needed. The number of each special item is listed per student (or student group). Total quantities of each item are to be determined according to class size. Some specialized equipment can be shared by several students (or student groups), e.g., balances, spectrophotometers, and leveling tanks. A complete list of specialized equipment and the experiment number in which each item is used (in parenthesis) is listed in Appendix D in this IM.

- **Answers** (and, on occasion, the solutions) to the Prelaboratory Assignment are provided.

- **Answers** (and, on occasion, the solutions) to the Laboratory Questions are also provided.

- A brief **Laboratory Quiz** (with answers) is developed for each experiment. Most often this would be completed after the experiment is performed.

The author welcomes all comments and criticisms from users of the laboratory manual and this IM. Suggestions for additions, corrections, and improvements to either/both manual(s) are invited.

The author is deeply indebted to his students, laboratory instructors, and stockroom personnel for their suggestions in refining the details that appear in this IM. The author greatly appreciates the continued support of Joan Kalkut, Associate Editor at Wiley, for her commitment toward excellence in the production of this laboratory package.

J. A. Beran   January, 1994
Box 161, Department of Chemistry
Texas A&M University-Kingsville
Kingsville, Texas 78363
USA

Contents

To the Laboratory Instructor

Teaching students in the general chemistry laboratory is a very challenging assignment. The students, having a cross-section of secondary school backgrounds and a wide range of laboratory experience, are pursuing a variety of undergraduate degrees. Therefore their purposes and aspirations in the course vary considerably. It is a challenge to turn students on to chemistry; however, it is much easier, more appropriate, and more fun in the laboratory.

Your assignment is to provide a meaningful chemistry laboratory experience for these students, most of whom are *not* chemistry majors, are in their first semester of college, are away from home for the first time, and are 17–18 years old.

GOALS FOR INSTRUCTION

1. Students must think scientifically: students must critically observe the chemical system, collect and analyze the data, make appropriate interpretations or calculations, draw a conclusion, and finally become confident in using their analysis for further interpretations.

2. Students must learn to be critical of their data and observations, be quantitative in their measurements and exact with the calculations, including the proper use of significant figures. (Review Laboratory Safety and Laboratory Guidelines, Part E in the manual.) The completion of the Report Sheet and Laboratory Questions becomes a "little" easier.

3. Students must develop good laboratory techniques. The extensive Laboratory Techniques section in the laboratory manual should be the most used part of the manual. In addition, you should constantly reinforce the benefits of acquiring and practicing the techniques during each laboratory period. An icon for each technique that should be used in the Experimental Procedure is placed at the closest position in the margin.

4. Students must acquire an appreciation of chemicals and practice proper laboratory safety guidelines. The Laboratory Safety and Laboratory Guidelines, Parts A–C in the manual, outline basic laboratory safety procedures for the general chemistry laboratory. Thoroughly review these laboratory safety procedures with students during the first laboratory period. Also *you* should review and be aware of "local" laboratory safety procedures.

5. Students must understand and appreciate the necessity of the proper disposal of test solutions and chemicals. (See Technique 4.) Educate yourself of the proper disposal policies of laboratory chemicals that are unique to your laboratories.

6. Students must understand the chemical principles for each experiment to ensure a greater appreciation of the principles, the data collection and analysis, and the observations encountered for each experiment.

INSTRUCTION ROUTINE

No more than 20 minutes should be taken from any laboratory period for instruction, which is preferably presented in a classroom. Quizzes should also be given during this 20 minutes period—students do *not* want to spend their entire laboratory session listening to you!

1. Take roll.

2. Require students to submit the Prelaboratory Assignment at the *beginning* of the laboratory period. Stress this procedure each laboratory session to ensure good laboratory preparation. We generally do not permit students to begin the experiment until this assignment is completed.

3. Review the grading and state your comments on the previous experiment.

4. Discuss the Objectives, the principles outlined in the Introduction, and give an overview of the Experimental Procedure. Various, significant techniques that are advantageous for a successful completion of the experiment should be stated. A technique icon is appropriately positioned in the margin of the Experimental Procedure.

5. Read and refer students to pages iv and 28 in the laboratory manual. Ultimately students are responsible for their own safety in the laboratory while conducting the experiment.

6. • Cite any cautions or other safety notes in the experiment. Each caution is noted with an icon, the international warning sign, in the margin of the Experimental Procedure.
   • Announce any changes that are to be made in the Experimental Procedure.
   • State the disposal procedure for the test chemicals. A disposal icon is appropriately positioned in the margin of the Experimental Procedure.

7. Assign the Laboratory Questions that are to be completed for the experiment.

8. Assign the experiment for the next laboratory session, if you have not already handed out a laboratory syllabus. Announce any changes or omissions that are appropriate for the next Prelaboratory Assignment.

INSTRUCTOR'S RESPONSIBILITIES

1. **Arrive at the laboratory** at least 15 minutes before the students to see if the laboratory benches and balance areas are clean, the stock chemicals are in adequate supply, the appropriate "Waste" container is present and empty, and the check-out equipment is available.

2. **Prepare for the lab.** Critically study the Instructor's Manual to anticipate student questions. This is your job, and you are expected to know more about the experiment than anyone else in that laboratory. Clarify uncertainties and ambiguities with your supervisor in advance of the day's experiment.

3. **Be assertive.** *You* must maintain proper student discipline. Enforce laboratory safety rules, proper laboratory techniques, proper disposal and cleanup procedures, and all other rules. (Review Laboratory Safety and Laboratory Guidelines, Parts C and D in the manual.) Relaxing the rules increases the probability of an accident caused by carelessness. If a student refuses to follow rules, remove the student from the laboratory. Taking a chance only increases the danger to the student, other students, and, most importantly, to *yourself*.

4. Above all, **be fair** to all students regardless of their personality, laboratory effort, or academic capabilities.

5. **Hand out** a laboratory syllabus for the laboratory program during the first laboratory period of the semester (term). This helps students to prepare.

PRELABORATORY ASSIGNMENT

A. True or False

1. F	6. T	11. T	16. T	21. F
2. F	7. F	12. T	17. F	22. F
3. T	8. F	13. F	18. F	23. T
4. T	9. F	14. F	19. T	24. F
5. T	10. T	15. T	20. T	25. T

B. Complete with the Correct Word(s), Phrase, or Value(s)

1. 2–3
2. water or glycerol
3. no water droplets
4. volumetric flask
5. ±0.01
6. zero
7. stirring rod
8. wash bottle
9. touches the beaker wall
10. disconnect the hose from the filter flask; turn off the water to the aspirator
11. an even; opposite
12. 20–40
13. no more than one-third
14. to minimize bumping and/or the formation of superheated steam
15. direct flame
16. 2 mL of 6 M HNO_3
17. black mark; just below
18. 30
19. left
20. eliminate air bubbles from the buret tip

The Laboratory and SI

INTRODUCTION

Be prepared! The first meeting between you and students is very important—first impressions are lasting. Set the proper "mood" for the lab and be specific in *all* of your instructions and expectations of student performance.

Spend time reading and reviewing the Preface and the section entitled "To the Laboratory Instructor" in the Instructor's Manual. Additionally, review the Preface in the Laboratory Manual. These sections inform you of what is available to you in helping with your instruction and supervision of the laboratory program.

Good preparation only helps your confidence for instruction and it establishes your credibility with students in the laboratory.

LECTURE OUTLINE

1. For your first day of instruction: identify the students (take roll), laboratory instructor (yourself), the laboratory section number, and each student's desk number. This information should be placed on the inside front cover of the manual.

2. Discuss the philosophy of the chemistry laboratory as suggested in the Introduction to Dry Lab 1.

3. Review the disclaimers that appears on pages iv and 28 of the laboratory manual.

4. Familiarize students with the laboratory manual: locate the Laboratory Safety and Laboratory Guidelines section, the list of Common Laboratory Equipment, the Laboratory Techniques, and the Appendices. Also review the format of each experiment: the Objectives, the Introduction, the Experimental Procedure, the Prelaboratory Assignment, the Report Sheet and the Laboratory Questions. Note that icons are used extensively in the Experimental Procedure to cite the Laboratory Techniques that are correspondingly appropriate.

5. Familiarize students with the laboratory: locate the safety equipment, the chemical stockroom, the balance room, and other physical facilities unique to your laboratories. *Require students to complete the inside front cover of the manual.*

6. Give a short lecture on laboratory safety and the disposal of chemicals. Note the use of the Caution icon in the Experimental Procedure. Use the Laboratory Safety section as a guide. Also discuss safety and disposal guidelines unique to your laboratories with your supervisor and the stockroom personnel.

7. Refer students to pages iv and 28. The disclaimer reminds students that they are responsible for their own safety while conducting experiments in the laboratory.

8. Review SI units and conversion factors. This section will take most of your laboratory time. To shorten the laboratory session you may assign only a portion of the Part C, the SI section.

9. Assign the experiment for the next laboratory period and determine what students will need to complete in advance. A syllabus for the laboratory covering the entire term is beneficial for students.

10. A quiz over laboratory safety and SI is suggested for the next laboratory period.

TEACHING HINTS

1. **Part A.** Assign each student to a laboratory station and issue equipment and glassware. A photograph of the suggested Common Laboratory Equipment and a check-in form are located on pages 6–7 of the Laboratory Manual.

2. To facilitate the check-in process, have all students place their drawer (or locker) equipment on the bench top; *you* identify an item and the students, in unison, return that item into the drawer and make the corresponding check (√) on the check-in form. If the student does not have the item, he/she can obtain it later from the stockroom, and *not* check the item on the check-in form.

You are to place your signature on the Report Sheet after the check-in process is completed.

3. Advise students to provide their own soap or detergent and paper towels (*not* the paper towels from the bathroom!).

4. **Part B.** Emphasize the importance of laboratory safety. Review with the students, in detail, Laboratory Safety and Laboratory Guidelines, pages 1–4. *Do not* neglect this discussion from your introductory remarks. You are to approve the completion of the inside front cover.

REPORT SHEET INFORMATION

B. Laboratory Safety

1. F, See Lab Safety, B.2
2. T
3. Hopefully true; if not, note its location
4. F, See Lab Safety, A.2
5. F, See Lab Safety, C.3
6. T
7. F, See Lab Safety, D.8
8. T
9. T
10. Hopefully true; if not, note its location
11. False!!!!
12. T
13. F, See Lab Safety, E.5
14. T
15. T
16. Hopefully true; if not, note its location
17. F, See Lab Safety, B.8
18. T

C. *Le Système International d'Unités* (SI Units)

1. Name the SI unit of measurement for:

   a. meter
   b. cubic meter
   c. kilogram
   d. kelvin
   e. pascal
   f. joule

2. Complete the following table.

   a. 3.3×10^{-12} g
   b. 7.6 mL
   c. 6.3 μamps
   d. 1.7×10^{3} watts
   e. 6.72 MJ
   f. 2.16×10^{-9} m
   g. 1.99 pg
   h. 4.21×10^{-1} L

3. Convert each of the following:

   a. $14.6 \text{ nm} \times \dfrac{10^{-9}\text{m}}{\text{nm}} \times \dfrac{\text{pm}}{10^{-12}\text{m}} = 14.6 \times 10^{3} \text{ pm} = 1.46 \times 10^{4} \text{ pm}$

   b. $250 \text{ mL} \times \dfrac{10^{-3}\text{ L}}{\text{mL}} \times \dfrac{\mu\text{L}}{10^{-6}\text{ L}} = 250 \times 10^{3} \text{ } \mu\text{L} = 2.50 \times 10^{5} \text{ } \mu\text{L}$

   c. $1.37 \text{ mg} \times \dfrac{10^{-3}\text{ g}}{\text{mg}} \times \dfrac{\text{kg}}{10^{3}\text{ g}} = 1.37 \times 10^{-6} \text{ kg}$

   d. $9.31 \times 10^{-3} \text{ g} \times \dfrac{\text{ng}}{10^{-9}\text{ g}} = 9.31 \times 10^{6} \text{ ng}$

4. Which unit expresses a larger quantity?

   a. $1 \text{ kg} > 1 \text{ lb}$
   b. $1 \text{ mL} > 1 \text{ }\mu\text{L}$
   c. $1 \,^{\circ}\text{C} > 1 \,^{\circ}\text{F}$
   d. $1 \text{ }\mu\text{g} > 1 \text{ ng}$
   e. $1 \text{ }\mu\text{m} > 1 \text{ nm}$

   f. $1 \text{ L} > 1 \text{ qt}$
   g. $1 \text{ kcal} > 1 \text{ J}$
   h. $1 \text{ atm} > 1 \text{ Pa}$
   i. $1 \text{ m} > 1 \text{ mm}$
   j. $1 \text{ oz} > 1 \text{ g}$

5. $0.50 \text{ in.} \times \dfrac{2.54 \text{ cm}}{\text{in.}} \times \dfrac{10^{-2}\text{m}}{\text{cm}} \times \dfrac{\text{mm}}{10^{-3}\text{m}} = 12.7 \text{ mm} \approx 13 \text{ mm}$

6. For 5 ft, 8 in.: $68 \text{ in.} \times \dfrac{2.54 \text{ cm}}{\text{in.}} = 173 \text{ cm}$

7. For 130 lb: $130 \text{ lb} \times \dfrac{453.6 \text{ g}}{\text{lb}} \times \dfrac{\text{kg}}{10^3 \text{g}} = 59.0 \text{ kg}$

8. 26 mi: $26 \text{ mi} \times \dfrac{5280 \text{ ft}}{\text{mi}} \times \dfrac{12 \text{ in.}}{\text{ft}} \times \dfrac{2.54 \text{ cm}}{\text{in.}} \times \dfrac{10^{-2} \text{ m}}{\text{cm}} \times \dfrac{\text{km}}{10^3 \text{ m}} = 41.8 \text{ km}$

   285 yd: $285 \text{ yd} \times \dfrac{36 \text{ in.}}{\text{yd}} \times \dfrac{2.54 \text{ cm}}{\text{in.}} \times \dfrac{10^{-2} \text{ m}}{\text{cm}} \times \dfrac{\text{km}}{10^3 \text{ m}} = 0.261 \text{ km}$

   Total distance: $41.8 \text{ km} + 0.261 \text{ km} = 42.1 \text{ km}$

9. For diameter = 1.6 cm, length = 15 cm: $V = \pi r^2 \ell = \pi(0.80 \text{ cm})^2 15 \text{ cm} = 30 \text{ cm}^3 = 30 \text{ mL}$

10. 4 ft 8 in. = 56 in.; $56 \text{ in.} \times \dfrac{2.54 \text{ cm}}{\text{in.}} = 142 \text{ cm}; 1.42 \text{ m}$

11. $12.0 \text{ fl. oz.} \times \dfrac{29.57 \text{ mL}}{\text{fl. oz.}} = 355 \text{ mL}$

12. $100 \text{ kJ} \times \dfrac{\text{kcal}}{4.184 \text{ kJ}} = 23.9 \text{ kcal}; \quad 100 \text{ kJ} \times \dfrac{10^3 \text{ J}}{\text{kJ}} \times \dfrac{\text{Btu}}{1054 \text{ J}} = 94.9 \text{ Btu}$

13. $50 \text{ tablets} \times \dfrac{325 \text{ mg}}{\text{tablet}} \times \dfrac{10^{-3}\text{g}}{\text{mg}} = 16 \text{ g}$

    $50 \text{ tablets} \times \dfrac{325 \text{ mg}}{\text{tablet}} \times \dfrac{10^{-3}\text{g}}{\text{mg}} \times \dfrac{\text{lb}}{453.6 \text{ g}} \times \dfrac{16 \text{ oz}}{\text{lb}} = 0.57 \text{ oz}$

Basic Laboratory Operations

INTRODUCTION

This is the first "experiment" that most students perform in the laboratory. Oftentimes, the "stone is cast" in this first laboratory session. Your presentation will determine, in large part, student expectations and goals. If you emphasize the importance of the laboratory, the techniques, and applied principles, you will have a well-run lab. Be a good, well-organized laboratory instructor. Be ready for this first, very important day!

In our laboratories, we stress the importance of good laboratory technique. At the conclusion of the Laboratory Techniques section in the laboratory manual, there is a Laboratory Technique Assignment. You may want to assign this as an in-lab or take-home assignment so that students become familiar with this section of the manual.

WORK ARRANGEMENT

Individuals. Begin half of the students on Part A; the other half on Part B.

TIME REQUIREMENT

2.5 hours

LECTURE OUTLINE

1. Follow the Instruction Routine outlined in "To the Laboratory Instructor".

2. Demonstrate with an explanation, the lighting of a Bunsen burner.

3. Discuss the proper use and care of balances. Balances are expensive, they are used extensively in this course and students must learn to handle them with respect.

4. Define density. It is an intensive and physical property of matter. Describe the procedure for its measurement.

5. Divide the students into two groups:
   Group I: Begin Part A
   Group II: Begin Part B

6. Class or group data are required for completing the data for the liquid unknown on the Report Sheet. Inform students how these data are to be collected.

7. Assign specific Laboratory Questions.

CAUTIONS & DISPOSAL

- Use tongs for holding the wire gauze and iron, copper, and aluminum wires in the flame.

- Where there is fire, there is danger.

- Do *not* pipet by mouth.

- Return the unknown solids for use in other laboratories. Dispose of the liquid unknowns in the "Waste Liquids" container.

TEACHING HINTS

1. **Part A.** All Bunsen burners are not the same; for example, some do not have a gas control valve. Advise students to adjust the Experimental Procedure accordingly.

   Assist students in the lighting and adjusting of the Bunsen burner. Make sure the tubing is attached to the gas outlet, *not* the water outlet! Remove combustible substances from the area near the Bunsen burner. Extinguish the flame when it is not in use.

2. **Part B.** Over-emphasize, if necessary, the care and operation of the balances. Students seem not to appreciate the quality of a balance (for some reason). Be strict while overseeing its operation *at all times*. Keep the balance area clean of all chemicals and glassware. See Technique 6.

3. **Part C.1.** Roll the metal in the water to remove air bubbles. Watch that students properly read a meniscus (see Technique 16A).

4. **Part C.2.** Supervise the use of the pipet and the proper pipetting technique—students are not to pipet with their mouths. (**Caution:** *keep liquid unknowns away from the Bunsen flame.*)

CHEMICALS REQUIRED	aluminum wire	5-10 cm
	copper wire	5-10 cm
	iron wire	5-10 cm

SUGGESTED UNKNOWNS

Issue a solid sample for Part C.1 and a liquid sample for Part C.3. Provide labeled containers for the return of the solid samples and a "Waste Liquids" container for the liquid samples.

Part C.1 Solid[a]	Density (g/cm^3)	Part C.3 Liquid[a]	Density (g/mL)
aluminum	2.70	methanol	0.791
copper	8.95	ethanol	0.789
iron (nails, *not* galvanized)	7.86	water	1.00
lead	11.34	1-propanol	0.804
nickel	8.90	toluene	0.867
silicon	2.42		
tin	7.28		
zinc	7.14		

[a]Volumes of 3–5 cm^3 of metal and 10 mL of liquid are needed.

SPECIAL EQUIPMENT	Bunsen burner	1	5-mL pipet and bulb	1
	match or striker	1	balance, ±0.01 g	1
	"Waste Liquids" container		balance, ±0.001 g	1

PRELABORATORY ASSIGNMENT

1. A luminous flame has a yellow color due to the burning of small carbon particles in the flame. A nonluminous flame is a blue flame where all of the fuel is being burned. The nonluminous flame produces higher temperatures.

2. Blue. A blue flame indicates a complete combustion of all carbon particles (and fuel).

3. Three distinct cones should appear in a properly adjusted Bunsen flame.

4. Two procedures:
   • holding a wire gauze in the flame, parallel to the burner barrel, to observe the flame profile
   • observing the temperature zones in which iron, copper, and aluminum metals melt

5. Triple-beam balance ±0.01 g
   Top-loading balance ±0.01 g and/or ±0.001 g

6. a. A meniscus is the horizontal surface of a liquid.
   b.

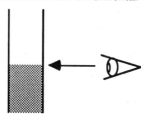

7. $\dfrac{8.248 \text{ g} - 6.684 \text{ g}}{2.00 \text{ mL}} = \dfrac{1.564 \text{ g}}{2.00 \text{ mL}} = 0.782 \text{ g/mL}$

8.   a.   touching the tip to the wall of the receiving vessel
     b.   nothing!
     c.   collapsed rubber pipet bulb
     d.   forefinger

LABORATORY QUESTIONS

1.   Solid A: volume displaced $= 1.00 \text{ g} \times \dfrac{\text{cm}^3}{2.70 \text{ g}} = 0.370 \text{ cm}^3$

    Solid B: volume displaced $= 1.00 \text{ g} \times \dfrac{\text{cm}^3}{3.87 \text{ g}} = 0.258 \text{ cm}^3$

    Solid A displaces a larger volume by $(0.370 - 0.258) \text{ g} = 0.112 \text{ cm}^3 = 0.112 \text{ mL}$

2. $16.44 \text{ g} \times \dfrac{\text{cm}^3}{11.35 \text{ g}} = 1.448 \text{ cm}^3 = 1.448 \text{ mL}$

    $1.448 \text{ mL} + 4.2 \text{ mL} = 5.6 \text{ mL}$

3.   *Lower*. The bubble displaces additional water, causing a larger recorded volume change for the metal; the larger its recorded volume, the *lower* its recorded density.

4.   *Less* than 2 mL. The liquid on the inner wall is a part of the calibrated 2-mL pipet.

LABORATORY QUIZ

1.   A properly adjusted Bunsen burner flame is nonluminous and has (one, two, three) distinct cones.   [Answer: 3]

2.   A 8.462-g metal bar changes the water level in a 50-mL graduated cylinder from 23.7 mL to 25.9 mL. Calculate the density of the metal.   [Answer: 3.85 g/cm^3]

3.   What is the criterion for clean glassware? [Answer: no water droplets cling to the wall]

4.   What is the fuel used for the flame in a Bunsen burner?
                          [Answer: generally, methane or natural gas]

5.   A 25.0-mL volume of a liquid was dispensed from a pipet. The mass of the liquid was determined to be 21.6 g. What is the density of the liquid?   [Answer: 0.864 g/mL]

Identification of a Compound: Physical Properties

INTRODUCTION	This is the first of two experiments that uses the properties of a substance for identification. In this experiment, physical properties are used to identify a compound; in the next experiment, the chemical properties are used. Density, solubility, melting point, and boiling point are used in this experiment to identify a compound.
WORK ARRANGEMENT	Individual unknowns; share the laboratory apparatus.
TIME REQUIREMENT	2.5 hours

LECTURE OUTLINE

1. Follow the Instruction Routine outlined in "To the Laboratory Instructor."

2. Review the disposal procedures for the experiment.

3. Demonstrate the procedure for filling the capillary tube used in Part C.1.

4. Approve the apparatus in Parts C and/or D prior to any heating.

5. Describe the procedure for the boiling point determination in Part D; there may be some confusion as to when the temperature should be read for recording of the boiling point.

CAUTIONS & DISPOSAL

- Most of the unknown liquids are flammable; be careful of the location of the Bunsen flame.

- Do not smell the unknown or the acetone and ethanol solvents.

- Dispose of the test solutions (Part A) and the liquid unknown (Part D) in a properly marked "Waste Organics" container.

- The temperature of the cooking oil in Part C may exceed 100°C.

- Mercury spills are possible with the use of the thermometers in this experiment. Discard mercury in the "Waste Mercury" container.

- Capillary tubes are to be discarded in the "Solid Wastes" container, *not* in the sink. Return rubber bands or $1/4$-inch pieces of tubing to the stockroom.

- Return the cooking oil to a "Used Oil" container.

TEACHING HINTS

1. **Part A.** Most unknown liquids are volatile, flammable organic compounds.

2. **Part B.** Review the proper technique for the reading of a meniscus and for pipetting a solution (Techniques 16A and 16B).

3. **Parts C and D.** A thermometer is likely to be broken. Clean up the mercury as best possible and place it in a "Waste Mercury" container. Dust the immediate area with powdered sulfur.

4. **Part C.** Supervise or demonstrate the proper technique for filling a capillary tube with a solid.

5. **Part D.** The boiling point of the liquid unknown is the temperature at which air bubbles cease to be evolved from the capillary tube...watch closely.

CHEMICALS REQUIRED		
95% ethanol	3–5 mL	
acetone	3–5 mL	
cyclohexane	3–5 mL	
boiling chips	2–3	

SUGGESTED UNKNOWNS

Unknowns are listed in the manual, Table 2.1. About 0.5 g of solid or 10 mL of liquid is required.

SPECIAL EQUIPMENT				
balance, ±0.01 g	1	110°C thermometer	1	
2-mL (or 5-mL grad.) pipet and bulb	1	Bunsen burner		
capillary melting point tube	2	ring stand and iron support rings		
360°C thermometer	1	"Solid Wastes" container		
cooking oil		"Used Oil" container		
rubber band or ¹/₄-inch piece of rubber tubing		"Waste Organics" container		
		"Waste Mercury" container		

PRELABORATORY ASSIGNMENT

1.
   a. water
   b. water
   c. gasoline
   d. gasoline
   e. water
   f. gasoline
   g. water

2.
   a. solubility: maximum mass of solute dissolved per fixed mass of solvent
   b. melting point: the temperature at which the solid and liquid phases coexist.
   c. boiling point: the temperature at which bubbles of liquid form spontaneously
   d. density: the mass of a substance per unit volume

3. *n*-hexane

4. The melting point is the most easily measurable difference between cyclohexane and cyclohexene.

5. The boiling point should be recorded when bubbles cease to escape from the capillary tube, *after* the heat is removed.

6. See Technique 16B.
   a. The pipet is rinsed with several small volumes of the liquid that is to be delivered.
   b. The liquid is drawn to above the calibrated mark on the pipet with a collapsed rubber bulb.
   c. Use the forefinger (not the thumb!) to dispense liquid down to the calibration mark on the pipet.
   d. Touch the pipet tip to the wall of the receiving flask, release the forefinger, and deliver the liquid from the pipet.

LABORATORY QUESTIONS

1. The greater the atmospheric pressure, the higher is the measured boiling point of the liquid.

2. *Too low.* A smaller volume will be delivered from the pipet, resulting in a lower measured mass. This will result in a *lower* reported density.

3.
   a. *Too high.* The temperature will exceed the boiling point if gas rapidly escapes the capillary tube.
   b. *Too low.* The more the liquid cools, the further it will rise into the capillary tube. The boiling point must be recorded *just* when the liquid enters the capillary tube.

LABORATORY QUIZ

1. Describe the apparatus and the observations required for measuring the boiling point of a liquid.

2. Explain why air bubbles escape the capillary tube (Figure 2.5) at the boiling point of a liquid.

3. Describe the procedure for measuring the density of a liquid.

4. Recalling the experience from Experiment 1, describe the procedure for measuring the density of a solid that is insoluble in water.

5. How does atmospheric pressure affect the boiling point of a liquid?
   [Answer: The greater the atmospheric pressure the higher is the boiling point of the liquid.]

6. List three measurable physical properties which can be used to identify a compound.
   [Answer: From this experiment, solubility, density, boiling point, and freezing point]

Experiment 3

Identification of a Compound: Chemical Properties

INTRODUCTION	In contrast to Experiment 2 where physical properties were used for identification, this experiment focuses on the chemical properties of a substance for identification. You may recognize this experiment as an open-ended version of an anion qual scheme; it is a good experiment for students to make observations and draw their own conclusions.
WORK ARRANGEMENT	Partners for the known compounds; individuals for the unknown.
TIME REQUIREMENT	2.5 hours

LECTURE OUTLINE

1. Follow the Instruction Routine outlined in "To the Laboratory Instructor."

2. Discuss, with examples, various observations that signify a chemical reaction:
   - gas evolved when an acid in placed on a carbonate salt
   - precipitate appears when calcium ion is mixed with a carbonate ion
   - heat is evolved in an acid-base neutralization reaction
   - a color change of the leaves occurs during the Fall season

3. Advise students of the procedure for filling out the "reaction matrix" on the Report Sheet.

4. **Part A.** Discuss the chemistry of each salt; explain how one anion can be identified in the presence of a number of anions.

5. Explain that once the observation of a chemical reaction is made, it should be recorded and that the purpose of subsequent testing is to characterize the reaction. For example, the OH^- ion precipitates with Mg^{2+}, but Mg^{2+} is soluble with the addition of $HCl(aq)$.

6. **Part B** is nearly open-ended. Students must rely entirely on observations for a determination of their unknown.

CAUTIONS & DISPOSAL

- None of the chemicals in this experiment are considered dangerous, but, if there is any contact with the skin, wash the skin immediately. You (or the student) will find out the next day if any silver nitrate has touched the skin!

- The salt solutions should be discarded in a "Waste Salts" container.

TEACHING HINTS

1. Students should use a clean 24-well plate and glassware; contamination of glassware can cause the appearance of "weird" precipitates!

2. **Part A.** Provide discussion of the observations and interpretations of the reaction mixtures.

3. We oftentimes issue an unknown to students at the *beginning* of the lab; by doing this, the students can place their unknown solution in wells A6–C6 and make comparisons as they proceed through Part A.1–3. This reduces the time for analysis and it eliminates the question "what color is this"?

REPORT SHEET
INFORMATION

Test	NaCl	Na_2CO_3	$MgSO_4$	NH_4Cl	H_2O
$AgNO_3$	p	p	(p)?	p	nr
NaOH	nr	nr	c	go	nr
HCl	nr	g	nr	nr	nr

CHEMICALS REQUIRED

Place the following in dropper bottles to minimize the waste of solutions.

Part A

0.1 M NaCl	1 mL	0.1 M $AgNO_3$	2 mL
0.1 M Na_2CO_3	1 mL	0.1 M NaOH	2 mL
0.1 M $MgSO_4$	1 mL	0.1 M HCl	2 mL
0.1 M NH_4Cl	1 mL		

Part B

A large number of solutes and test reagents can be selected to provide the problem required in Part B. Following are only suggestions:

Set 1

0.1 M HCl	1 mL	phenolphthalein	2 mL
0.1 M HNO_3	1 mL	0.1 M Na_2CO_3	2 mL
0.1 M NaOH	1 mL	0.1 M $AgNO_3$	2 mL
satd $Ca(OH)_2$	1 mL		
0.1 M KNO_3	1 mL		

Set 2

0.1 M Na_3PO_4	1 mL	0.1 M $Ba(NO_3)_2$	2 mL
0.1 M Na_2CO_3	1 mL	0.1 M HNO_3	2 mL
0.1 M Na_2SO_4	1 mL	0.1 M $CuSO_4$	2 mL
0.1 M Na_2S	1 mL		
0.1 M NaCl	1 mL		

Set 3

0.1 M HCl	1 mL	phenolphthalein	2 mL
0.1 M NaOH	1 mL	0.1 M Na_2CO_3	2 mL
0.1 M Na_2SO_4	1 mL	0.1 M $Ba(NO_3)_2$	2 mL
0.1 M NH_3	1 mL		
0.1 M H_3PO_4	1 mL		
or			
satd $Ca(OH)_2$	1 mL		
0.1 M $CuSO_4$	1 mL		
0.1 M HNO_3	1 mL		
0.1 M CH_3NH_2Cl	1 mL		

SPECIAL EQUIPMENT

24-well plate and Beral pipets	1
"Waste Salts" container	

PRELABORATORY ASSIGNMENT	1.	a.	$NaNO_3$	soluble
		b.	$Mg(OH)_2$	insoluble
		c.	NH_4NO_3	soluble
		d.	$MgCl_2$	soluble
		e.	Ag_2CO_3	insoluble
		f.	$AgCl$	insoluble

2. Open-ended response. Suggest carbonated vs. noncarbonated beverages as first separation, juices vs. nonjuices, natural liquids vs. commercial (or processed) liquids

3. test tube 1   .   potassium iodide
    test tube 2   sodium sulfide
    test tube 3   silver nitrate

LABORATORY QUESTIONS

1. a.   test tube 1   sodium carbonate
          test tube 2   hydrochloric acid
          test tube 3   silver nitrate
   b.   A precipitate of silver chloride would form. See the solubility rules in the manual, Appendix G.

2. a.   carbon dioxide gas
   b.   silver chloride precipitate
   c.   magnesium hydroxide precipitate
   d.   ammonia gas

3. a. $AgCl$
   b. $BaSO_4$
   c.   (1) add HCl and separate, (2) add H_2SO_4 and separate; (3) add excess NH_3 to the supernatant

LABORATORY QUIZ

1. Cite two observations that indicate the occurrence of a chemical reaction.
[Answer: see p. 59 of the laboratory manual]

2. A mixture of sodium sulfate and ammonium chloride produces no observable result; however, the mixture of sodium hydroxide and ammonium chloride produces a detectable odor. What is the odor? [Answer: NH_3 gas]

3. A mixture of lead nitrate and ammonium acetate produces no observable result; however, the mixture of lead nitrate and ammonium sulfate produces a white precipitate. What is the precipitate? [Answer: lead sulfate]

4. A mixture of sodium carbonate and sodium hydroxide produces no observable result; however, the mixture of sodium carbonate and barium hydroxide produces a white precipitate. What is the precipitate? [Answer: barium carbonate]

5. Describe the technique for detecting an odor. [Answer: See Technique 17A]

Experiment 4
Paper Chromatography

INTRODUCTION	The separation of mixtures, whether they be homogeneous or heterogeneous, is one of the challenges of a chemist. Many separation techniques are used, some of which are familiar; others, such as chromatography, are new and students find them interesting.
WORK ARRANGEMENT	Partners.
TIME REQUIREMENT	2.5 hours

LECTURE OUTLINE

1. Follow the Instruction Routine outlined in "To the Laboratory Instructor."

2. Inform students that chromatography is a unique separation technique that takes advantage of the adsorptive properties of a substance. Paper chromatography is only one type of chromatography—others include gas-solid, liquid-solid, thin-layer, column, . . .

3. The stationary phase in this experiment is the "paper;" the mobile phase is the transition metal ions dissolved in the eluting solution of acetone and hydrochloric acid.

4. Discuss the techniques of spotting the chromatographic paper (Part B.1) and of the careful positioning of the spots on the chromatographic paper.

5. Each component (transition metal ion) of the mobile phase has an R_f (ratio of fronts) factor (Equation 4.1). Explain the R_f factor with an example.

6. The R_f factor for each transition metal ion is measured with a known solution; the R_f factor for each band in the unknown is also "developed" in the same chromatogram. A comparison of the R_f factors in the unknown with the known solution determines the composition of the unknown mixture.

CAUTIONS & DISPOSAL

- Acetone is flammable; keep all flames extinguished in the lab—no need for a flame in this experiment.

- 6 M HCl is used in Part A.1—be careful!

- **Part A.2.** Conc NH_3 is an irritant to the skin and the respiratory system.

- The eluting solution should be discarded in a "Waste Organics" container at the conclusion of the experiment.

TEACHING HINTS

1. **Part A.1.** The eluting solution should not wet the wall of the developing chamber—this also means that the beaker wall must be dry.

2. **Part A.3.** Only a very small drop of solution is needed to develop a chromatogram—a microdrop is sufficient. Practice spotting a piece of filter paper before spotting the chromatographic paper.

3. **Part A.4.** Handle only the top of the chromatographic paper with the hands; keep the paper free of benchtop contaminants.

4. **Part B.1.** The microdrop of each solution spotted on the chromatographic paper must be *above* the eluting solution in the beaker.

5. **Part B.3.** Be sure the developing chamber is covered with plastic wrap during the development of the chromatogram. Remove the chromatogram from the developing chamber when the solvent front is 1.5 cm from the top.

6. **Part C.2.** The identification of the bands with conc NH_3 vapors in the ammonia chamber requires a time judgment—this amplification technique requires patience.

7. **Part C.3** is optional, but the procedure does enhance the location of the transition metal ion band. If the band is already identified, there is no need for enhancement.

8. **Part C.4.** The R_f value should be determined from the center of the band. Identification of the transition metal ion can also be identified by its color.

CHEMICALS REQUIRED		
	0.1 M $Co(NO_3)_2$ in 0.1 M HNO_3	dropper bottle
	0.1 M $Cu(NO_3)_2$ in 0.1 M HNO_3	dropper bottle
	1 M $Ni(NO_3)_2$ in 0.1 M HNO_3	dropper bottle
	0.1 M $FeNH_4(SO_4)_2$ in 0.1 M HNO_3	dropper bottle
	1 M $Mn(NO_3)_2$ in 0.1 M HNO_3	dropper bottle
	acetone	9 mL
	6 M HCl	1 mL
	conc NH_3	5 mL

The unknown samples U1, U2, and U3 are to consist of combination of the same 5 cations. Generally 1 cation is the U1 sample, 2 cations are in the U2 sample, and 3 cations are in the U3 sample.

OPTIONAL CHEMICALS FOR PART C.3		
	0.1 M $NaBiO_3$	dropper bottle
	0.2 M KSCN	dropper bottle
	satd KSCN in acetone	dropper bottle
	0.1 M or 1% dimethylglyoxime	dropper bottle
	0.2 M $K_4[Fe(CN)_6]$	dropper bottle

Any combination of the five cations can be used for the three unknowns; the cations Fe^{3+} and Co^{2+} are not *clearly* separated in the experiment.

SPECIAL EQUIPMENT		
	600-mL beaker	1
	plastic wrap and rubber band	
	1000-mL beaker	1
	capillary tubes	8–13
	metric ruler	
	paper clip or stapler	
	chromatographic paper, 4" x 8"	
	(suggest Whatman No. 1)	1
	"Waste Organics" container	

PRELABORATORY ASSIGNMENT

1. a. A solvent front is the leading edge of the solvent as it is adsorbed and moves upward on the stationary phase (the chromatographic paper).
   b. The stationary phase is the adsorptive material to which components in the mobile phase adhere.
   c. The mobile phase is the solution that contains the components of the mixture.
   d. $R_f = \dfrac{\text{distance from origin to final position of component in mobile phase}}{\text{distance from origin to the solvent front}}$

2. The developing chamber is covered with plastic wrap to prevent the chromatographic paper from drying during the development of the chromatogram.

3. Top: $R_f = \dfrac{2.0 \text{ cm}}{8.2 \text{ cm}} = 0.24$

    Middle: $R_f = \dfrac{7.3 \text{ cm}}{8.2 \text{ cm}} = 0.89$

    Bottom: $R_f = \dfrac{5.0 \text{ cm}}{8.2 \text{ cm}} = 0.61$

4. Color and R_f value.

5. $R_f = \dfrac{27 \text{ mm}}{88 \text{ mm}} = 0.31$; $R_f \times 102 \text{ mm} = 32 \text{ mm}$

LABORATORY QUESTIONS

1. *Larger R_f values.* The R_f values would be greater because the solvent front would evaporate (near the top of the chamber), while the components continue to move.

2. The eluting solution would cause the band to diffuse, thus making it more difficult to determine the R_f value of the component in the mobile phase.

3. A ballpoint or ink pen contain ink that is a mixture of dyes. These dyes may also elute onto the chromatographic paper and mix with components originally in the mobile phase, thus contaminating the chromatogram.

4. Change eluting solutions.

*5. A higher mole ratio of HCl to acetone increases the polarity of the eluting solution, thus changing the R_f factors for the cations.

LABORATORY QUIZ

1. The R_f value of a dye in an ink is 0.127; the R_f of a second dye is 0.685. Which dye has a greater affinity for the stationary phase? Which dye progresses further along the stationary phase?
    [Answer: the first dye has the greater affinity for the stationary phase; the second dye progresses further along the stationary phase]

2. Can paper chromatography be used to separate the components of a heterogeneous mixture? Explain.

3. The larger the R_f value the (greater, lesser) its movement along the stationary phase.
    [Answer: Greater]

4. a. Characterize (or describe) the mobile phase in paper chromatography.
    b. Characterize (or describe) the stationary phase in paper chromatography.

The Chemistry of Copper

INTRODUCTION
Copper metal is cycled through a number of different copper compounds and then regenerated. As the original copper sample progresses through a series of chemical reactions, different colors and chemical forms of the copper ion are produced. The percent recovery of copper is finally determined. The experiment generates interest and intrigue.

WORK ARRANGEMENT
Partners. Students gain a better understanding and appreciation of the chemical changes if they have a chance to discuss their observations.

TIME REQUIREMENT
3 hours

Each reaction is categorized. Notice that in the Introduction we initially avoid the use of the terms, oxidation-reduction and acid-base reactions. We refer to these reactions as types of displacement reactions, but then later indicate that they can also be called oxidation-reduction and acid-base reactions.

LECTURE OUTLINE
1. Follow the Instruction Routine outlined in "To the Laboratory Instructor."

2. Review, using appropriately balanced equations, the cycle of chemical reactions that are performed in the experiment.

3. Introduce students, with examples, to single and double displacement reactions and to decomposition reactions.

4. The precautions for heating a test tube with a cool flame (Part C.1) should be emphasized before the experiment begins.

5. Advise students to follow the chemistry of the reactions with Equations 5.1–5.6 as they proceed through the Experimental Procedure.

6. Since the percent recovery of the copper is calculated at the conclusion of the experiment, you may want to discuss this calculation.

CAUTIONS & DISPOSAL
- **Parts C.1 and D.1.** Heating a small test tube with a direct flame (Technique 13A) should be done carefully. Caution students that the direct heating should be done carefully and with a "cool" Bunsen flame!

- Conc HNO_3, $NO_2(g)$, 6 *M* NaOH, 6 *M* H_2SO_4 are produced and/or used in this experiment. Caution students to their handling and exposure.

- $NO_2(g)$ concentrations, while small, can be discomforting. A fume hood is recommended for use in Part A.2.

- Dispose of the recovered copper metal in the "Waste Solids" container.

TEACHING HINTS
1. The equations requested on the Report Sheet are listed in the Introduction.

2. **Part A.1.** A larger test tube can be used for the reactions of copper, depending on the sleeve size of your laboratory's centrifuges.

3. **Part A.** Do not use more than 0.02 g of Cu wool. An excess of copper complicates the experiment. Students should not (and probably won't) inhale the $NO_2(g)$. The use of a fume hood is recommended.

4. **Parts B and E.** The technique for separating a liquid from a solid by centrifugation is used. Make sure students properly operate the centrifuge.

5. **Parts C and D.** *CAUTION*. Heating a small test tube over a direct flame may cause its contents to be ejected (Technique 13a). This procedure is to be done *carefully* and with a "cool" flame.

6. **Part E.** Make sure that all of the Cu^{2+} is displaced by the Mg metal and that an excess of H_2SO_4 is present so that the solution remains clear after the displacement. Do *not* overheat the Cu metal—in Part E.3, it says "Gently dry...over a 'cool' flame."

7. Instructor approval is required at each stage of the copper cycle.

CHEMICALS REQUIRED		
copper wool	0.02 g	
conc HNO_3 (dropper bottle)	1 mL	
6 M NaOH	1.5 mL	
6 M H_2SO_4	2 mL	
Mg ribbon	5-10 cm	

SPECIAL EQUIPMENT			
balance (±0.001 g)			Bunsen burner
steel wool			centrifuge
dropping (or Beral) pipets		5	"Waste Solids" container
fume hood			

PRELABORATORY ASSIGNMENT

1. a. Decant. To pour off the liquid/solution above a precipitate
   b. Supernatant. The liquid/solution above the precipitate in a test tube, beaker, or flask.
   c. Balancing the centrifuge. To place a test tube containing a liquid of equal volume to that of the test solution directly across the spindle of the centrifuge.
   d. Cool flame. A Bunsen flame that is adjusted with a low fuel supply.

2. a. 45°
   b. 20–40 s
   c. a minimum of two

3.
Formula	Name	Color
HNO_3	nitric acid	colorless
H_2SO_4	sulfuric acid	colorless
CuO	copper(II) oxide	black
$Cu(OH)_2$	copper(II) hydroxide	light blue
$CuSO_4$	copper(II) sulfate	blue
$Cu(NO_3)_2$	copper(II) nitrate	blue
NaOH	sodium hydroxide	colorless
$NaNO_3$	sodium nitrate	colorless
NO_2	nitrogen dioxide	red-brown
$MgSO_4$	magnesium sulfate	colorless

4. $0.02 \text{ g} \times \dfrac{\text{mol}}{63.54 \text{ g Cu}} \times \dfrac{4 \text{ mol } HNO_3}{\text{mol Cu}} \times \dfrac{\text{L}}{16 \text{ mol } HNO_3} \times \dfrac{\text{mL}}{10^{-3} \text{ L}} \times \dfrac{20 \text{ drops}}{\text{mL}}$
   = 1.6 drops

LABORATORY QUESTIONS

1. *Low.* If the solution remains blue, some Cu^{2+} remains in solution, which is subsequently discarded (Part C) and therefore cannot be recovered in Part E.

2. *High.* The Mg metal, if not completely dissolved, will be measured along with the Cu metal, producing an error showing excessive percent recovery of copper.

3. a. If the reported % recovery of copper is greater than 100%, then
   • the copper sample may not be dry
   • some Mg metal may remain unreacted in the copper sample
   • some CuO may form if the recovered Cu metal is heated too strongly in Part E.3
   b. If the reported % recovery of copper is less that 100%, then
   • copper ion may have been lost in the precipitation of $Cu(OH)_2$, Parts B and C
   • copper may not have been recovered from the solution in Part E

1. The color of the copper(II) ion is _____[Answer: sky blue]_____.

2. Write a balanced equation for the displacement of copper(II) ion from solution with magnesium metal.   [Answer: $Cu^{2+}(aq) + Mg(s) \rightarrow Cu(s) + Mg^{2+}(aq)$]

3. When heating a test tube directly with a *cool* flame, the test tube should be held at a _[Answer: 45°]__ angle.

4. Complete the table.

Formula	Name
HNO_3	_____
_____	copper(II) hydroxide
NaOH	_____
_____	sulfuric acid
NO_2	_____

5. A 0.0233-g sample of copper metal was cycled through a series of chemical reactions and then recovered. The recovered copper metal had a mass of 0.0194 g. What is the percent recovery of copper metal?   [Answer: 83.3%]

Formula of a Hydrate

INTRODUCTION	A gravimetric analysis is used to determine the mole ratio of an anhydrous salt to its volatile hydrated water molecules in a hydrated salt. Students use the mole concept for calculations.
WORK ARRANGEMENT	Individuals. Three trials required.
TIME REQUIRED	3 hours (including calculations)

LECTURE OUTLINE

1. Follow the Instruction Routine outlined in "To the Laboratory Instructor."

2. Impress on students the quantitative nature of this determination. Students should only use crucible tongs when handling the crucible and cover once the crucible and lid have been fired (see Figure 6.2).

3. A sample calculation for showing the mole ratio of an anhydrous salt to its waters of hydration may be helpful (for example, Prelab Question 1).

4. Three analyses require you and the students to be well prepared. Time spent in the balance area should not be wasted.

5. Advise students not to overheat the crucible containing the salt. Excessive heating may cause the anhydrous salt to decompose.

6. Repeated mass measurements of the crucible after the water has been removed should be within ± 0.010 g.

CAUTIONS & DISPOSAL

- Hot and cold crucibles look the same! *Do not touch!* Burned fingertips are common to those students who are careless or do not wish to develop the technique of handling the crucible and lid with crucible tongs.

- Do not place a hot crucible on the balance or on the benchtop.

- Anhydrous salts should be discarded in the "Waste Solids" container.

TEACHING HINTS

1. **Part A.1.** To remove difficult stains and/or carbon deposits from crucibles, add 2–3 mL of 6 M HNO_3, transfer to a fume hood, and heat to dryness. This procedure should be used with discretion.

2. **Part A.1.** Only *cool* crucibles are to be placed on the balance.

3. **Part B.1.** Heat the crucibles containing the sample, slowly at first and then gradually intensify the heat.

4. **Part B.2.** Do *not* overheat the salt; this avoids its decomposition.

5. Results are <3%.

CHEMICALS REQUIRED		
6 M HNO_3	2–5 mL	
1 M HCl	2–3 mL	

A 3-g sample of a hydrated salt is required *per trial*; for three trials a mass of about 10 g is required per student.

$BaCl_2 \cdot 2H_2O$	(loses 2 H_2O at 113°C)
$Na_2CO_3 \cdot H_2O$	(loses 1 H_2O at 100°C)
$ZnSO_4 \cdot 7H_2O$	(loses 7 H_2O at 280°C)
$CaSO_4 \cdot 2H_2O$	(loses 1 $^1/_2$ H_2O at 128°C, 2 H_2O at 163°C)
$CuSO_4 \cdot 5H_2O$	(loses 4 H_2O at 110°C, 5 H_2O at 150°C)
$MgSO_4 \cdot 7H_2O$	(loses 6 H_2O at 150°C, 7 H_2O at 200°C)

**SPECIAL
EQUIPMENT**

crucible tongs		Bunsen burner
crucible and lid	3	ring stand and iron support ring
clay triangle	1	desiccator or desicooler (optional)
balances, ±0.001 g		"Waste Solids" container

**PRELABORATORY
ASSIGNMENT**

1. a. 20.93% H_2O
    b. 0.03729 mol H_2O and 0.01865 mol $CaSO_4$
    c. $CaSO_4 \cdot 2H_2O$

2. If the hydrated salt is heated too strongly, the anhydrous salt itself may decompose. The decomposition products, if volatile, would produce a mole ratio of salt to water that is too low.

3. Initially heat the hydrated salt for 10 minutes, followed by repeated 2 minute heatings until successive mass measurements are ± 0.010 g.

4. Crucible tongs are used to handle crucibles and lids to minimize the transfer of contaminants. Using fingers to handle crucibles and lids transfers body oil and also increases the possibility of their being burned, as hot and cold crucibles look the same.

5. a. Firing the crucible means to heat the crucible to high temperatures (dull red) to remove any volatile impurities from the crucible.
    b. The lid should be set so that $^2/_3$ of the contents of the crucible is exposed to the atmosphere.

**LABORATORY
QUESTIONS**

1. *Too high.* The volatile impurities will be calculated as part of the mass loss of water from the hydrated salt.

2. *Too high.* The mass loss due to the decomposition of the anhydrous salt will be considered as water loss.

3. *High.* The remaining solid will be considered salt, resulting in a smaller mass of water removed and a higher salt to water mole ratio.

4. $CaCl_2$ adsorbs water vapor from the atmosphere into its crystalline structure, forming a hydrated salt.

5. *Too low.* The oil from the lab bench would be considered a part of the hydrate salt. In the heating process, the waters of hydration and the oil would be volatilized, leaving the mass of the remaining anhydrous salt as being too low.

LABORATORY QUIZ

1. Write the formula of the dihydrate of calcium sulfate.   [Answer: $CaSO_4 \cdot 2H_2O$]

2. How many grams of water are in 5.00 g of $CuSO_4 \cdot 5H_2O$?   [Answer: 1.80 g]

3. A hydrated salt is heated to determine the mole ratio of salt to water.
    a. If all of the hydrated water molecules are not removed, is the calculated mole ratio of salt to water high or low? Explain.   [Answer: high]
    b. If the hydrated salt is overheated to the extent that the salt decomposes into a volatile product, is the calculated mole ratio of salt to water high or low? [Answer: $\frac{salt}{water}$ is *low*]

4. In an experiment, the hydrated salt was heated according to the accepted experimental procedure; however, the sample was allowed to cool without the crucible being covered. Will the salt to water mole ratio be reported high or low? Explain.

    [Answer: $\dfrac{\text{salt}}{\text{water}}$ is *low*, water is absorbed by the anhydrous salt from the atmosphere]

5. A 2.789-g sample of a hydrated sodium carbonate salt was heated in a crucible. After heating, a reproducible mass of 2.384 g of anhydrous sodium carbonate was obtained. What is the formula of the hydrated salt? Molar mass of Na_2CO_3 = 106 g/mol.

    [Answer: $Na_2CO_3 \cdot H_2O$]

Experiment 7
Empirical Formulas

INTRODUCTION

This second gravimetric analysis experiment (see Experiment 6) establishes the empirical formulas of two compounds; one by a synthesis reaction, the second from a decomposition reaction. Good data are obtained for determining the empirical formula for magnesium oxide. The decomposition reaction for determining the empirical formula for calcium carbonate requires the student to "think" more about the analysis of the data.

WORK ARRANGEMENT

Individuals. Three trials.

TIME REQUIRED

2 hours for either the magnesium oxide or the calcium carbonate analysis

LECTURE OUTLINE

1. Follow the Instruction Routine outlined in "To the Laboratory Instructor."

2. Review the definition of empirical formula and how empirical formulas are determined experimentally.

3. Divide the laboratory students into two groups. Assign one group to the magnesium-oxygen synthesis, Part B, and the other group (the more advanced students) to the calcium carbonate decomposition, Part C.

4. Successive mass measurements of the final product should be ±1%

CAUTIONS & DISPOSAL

- Hot and cool crucibles look the same! Handle crucible and lid with crucible tongs after the first heating in Part A.1.

- The crucible and lid must be cooled to room temperature before placing them on the balance.

- The magnesium oxide and the calcium oxide should be properly discarded. Discuss the chemistry of hydrochloric acid for the proper disposal of the products.

TEACHING HINTS

1. Deviations from the expected empirical formulas are due to an incomplete reaction. Encourage students to report their nonstoichiometric formula from their data analysis. In other words, report their data as they have it, *not* what they think it ought to be.

2. The balance area will be a busy place! Oversee the proper operation of the balance and the maintenance of a clean balance area.

3. **Part A.3.** Advise the students of the availability and use of a desiccator.

4. **Part B.1.** Clean the Mg metal with steel wool. A 0.3-g sample of Mg provides about 0.5 g MgO; this is sufficient for the experiment. A 0.3-g sample of Mg is a 30 cm strip of Mg ribbon. Avoid the excessive use of Mg metal.

5. **Part B.2.** Heat the Mg ribbon slowly with the lid in position (see Figure 7.1).

6. **Part B.4.** Strive for ±1% reproducibility in successive mass measurements of the product.

7. **Part C.** Review the comments for Part B.

CHEMICALS REQUIRED

6 M HNO$_3$	2-5 mL
magnesium ribbon	0.7 g
calcium carbonate	2 g
1 M HCl (for cleanup)	2–3 mL

| SPECIAL EQUIPMENT | steel wool
crucible tongs
crucible and lid
balance (±0.001 g) | 3 | clay triangle
ring stand and iron support ring
Bunsen burner
desiccator or desicooler (optional) | 1 |

PRELABORATORY ASSIGNMENT

1. Cr_2O_3

2. Au_2O_3

3. The mass of oxygen is determined from a difference of the mass of product in the crucible after heating and the initial mass of the magnesium metal.

4. a. First, hot crucibles burn fingers; second, fingers leave oily deposits on the crucible and lid which have mass and thus affect the mass measurements, especially when sensitive balances are used.
   b. Cooling a crucible and lid in a desiccator reduces the amount of moisture that condenses on the crucible, lid, and sample during the cooling process.

LABORATORY QUESTIONS

1. In Part B, the magnesium metal forms a powder. In Part C, the appearance of the calcium carbonate changes.

2. a. *No effect.* The mass of the magnesium is unaffected because its mass is determined at the beginning of the experiment.
   b. *Less.* The difference between mass of the magnesium oxide and the initial mass of the magnesium metal is less because of some loss (volatilization) of the magnesium oxide, resulting in a lower recorded mass of oxygen in the product.

3. The mole ratio of CaO to CO_2 will be high because the $CaCO_3$ will not have been fully decomposed, leaving a mass in the crucible that will be assumed to be CaO.

4. *High.* The presence of the Mg_3N_2 will imply less oxygen combining with the fixed mass of magnesium resulting in a reported higher magnesium to oxygen mass ratio.

5. a. Ag_2S
   b. 87.0% Ag, 13.0 % S

LABORATORY QUIZ

1. a. A 0.0430-g sample of silver metal is heated with sulfur forming 0.0494 g of silver sulfide. Calculate the empirical formula of silver sulfide.
   b. What is the percent by mass of silver and sulfur in the sample?
   [Answer: a. Ag_2S; b. 87.0% Ag, 13.0 % S]

2. If the Mg ribbon is not cleaned to remove its oxide coating before its initial mass measurement, will the reported Mg to O mole ratio be high or low? Explain.
   [Answer: *high*, less oxygen reacts than what is expected]

3. Why is it best to use a desiccator for the cooling of the crucible and lid (and sample) after an initial firing and after all subsequent heatings, instead of the laboratory bench?
   [See answer to Prelab Question 4b]

Limiting Reactant

INTRODUCTION

The concept of the limiting reactant is always discussed near the end of the stoichiometry chapter in most textbooks. Therefore, it is appropriate for students to determine a limiting reactant and the percent composition of a salt mixture. In this experiment the reaction of $BaCl_2 \cdot 2H_2O$ with $Na_3PO_4 \cdot 12H_2O$ is used to illustrate the limiting reactant concept.

WORK ARRANGEMENT

Individuals. Two trials.

TIME REQUIREMENT

2 hours for two experimental trials, $1/2$ hour for drying in an oven, and $1/2$ hour for calculations.

LECTURE OUTLINE

1. Follow the Instruction Routine outlined in "To the Laboratory Instructor."

2. Review the principles of the limiting reactant concept *and* the percent composition. Even though limiting reactant calculations may have been discussed in lecture, an explanation of "how" the analysis is to be completed in the laboratory should be explained.

3. Discuss Parts A and B. Students often do not clearly understand the limiting reactant *test* in Part B.

4. Demonstrate the technique for folding filter paper and "sealing" it into a glass funnel for the gravity filtration procedure.

5. Review the calculations, either before or after the experiment. Explain that the hydrated water molecules of the $BaCl_2$ and Na_3PO_4 salts are included in the calculations. One or two days may be allowed to complete these calculations; however, if the Report Sheet leaves the laboratory, initial it to avoid any "juggling" of data.

CAUTIONS & DISPOSAL

• The experiment is relatively safe with regard to the chemicals used.

• A "Waste Solids" container should be available for the disposal of the barium phosphate precipitate at the conclusion of the experiment.

TEACHING HINTS

1. Survey the amount of oven space that is available for drying the precipitate—2 samples per student. If adequate oven space is not available, the samples must be air-dried in student desks.

2. **Part A.1.** Measure the 200 mL of water to within ±0.2 mL. Keep the balances and balance areas *clean*.

3. **Part A.2.** The purpose of the "digesting" the precipitate is to coagulate and reduce the loss of precipitate in the filtering process.

4. **Part A.3.** Inspect the filtering apparatus before the filtration begins in Part A.5. Make certain the filter paper is properly sealed in the funnel. A vacuum filtering procedure can be used, however, more of the precipitate will pass through the funnel because of the fine crystalline size of the precipitate.

5. **Part A.4.** The 50-mL decanted volumes should be measured to within ±0.2 mL. Stress quantitative techniques for reading a meniscus, T.16a.

6. **Part A.5.** Oversee the filtering of the $Ba_3(PO_4)_2$. Filter the suspension while hot and through fine porosity filter paper.

7. **Part A.5.** Even though fine porosity filter paper is recommended, some of the finely divided $Ba_3(PO_4)_2$ precipitate will pass through the filter paper, causing the filtrate to appear cloudy. Students are to comment on this inherent error in Laboratory Question, No. 1.

8. **Part A.6.** The precipitate can be air-dried or oven-dried, whichever is convenient in your laboratory.

9. **Part B.** Some explanation of the limiting reactant test may be necessary.

CHEMICALS REQUIRED

0.5 M $BaCl_2$ (dropper bottle)   1 mL
0.5 M Na_3PO_4 (dropper bottle)   1 mL

In addition, place several of these labeled dropper bottles of each solution in the laboratory.

SUGGESTED UNKNOWNS

Stir the mixtures thoroughly to achieve the *best possible* homogeneous mixture. Prepare 3 grams of an unknown sample per student.

Sample Salt Mixture	$Na_3PO_4 \cdot 12H_2O$(powder)	$BaCl_2 \cdot 2H_2O$(crystal)
1 mol Na_3PO_4/3 mol $BaCl_2$	38.0 g	73.2 g
1 mol Na_3PO_4/2 mol $BaCl_2$	38.0 g	48.8 g
2 mol Na_3PO_4/1 mol $BaCl_2$	76.0 g	24.4 g
3 mol Na_3PO_4/1 mol $BaCl_2$	114.0 g	24.4 g

SPECIAL EQUIPMENT

watchglass to cover 400-mL beaker	1	thermometer, 110°C
weighing paper	2	Bunsen burner
balance (±0.001 g)	1	ring stand and iron support ring
filter paper (fine porosity,		rubber policemen
No. 42 Whatman or Fisher*brand* Q2)	2	drying oven (optional)

PRELABORATORY ASSIGNMENT

1. a. Digesting the precipitate means to increase the crystalline size of the precipitate with heat.
   b. Warm the precipitate to an elevated temperature and maintain the temperature for an extended time for the purpose of increasing the crystalline size of the precipitate.

2. a. The mass difference between that of the filter paper plus the $Ba_3(PO_4)_2$ and that of the filter paper determines the mass of $Ba_3(PO_4)_2$.
   b. Add 2 drops of 0.5 M $BaCl_2$ to 50 mL of the supernatant from Part A.

3. 0.245 g $Na_3PO_4 \cdot 12H_2O$

4. $BaCl_2 \cdot 2H_2O$ is the limiting reactant producing 0.433 g $Ba_3(PO_4)_2$.

5. Fine porosity filter paper is required for the filtering of $Ba_3(PO_4)_2$ because $Ba_3(PO_4)_2$ is a very finely divided precipitate.

LABORATORY QUESTIONS

1. Percent limiting reactant is *low*. The mass of the limiting reactant directly determines the mass of $Ba_3(PO_4)_2$; thus if $Ba_3(PO_4)_2$ is lost in the filtering process, the mass of the limiting reactant is also reported as being low.

2. a. 0.104 mg $Ba_3(PO_4)_2$, 1.72×10^{-7} mol $Ba_3(PO_4)_2$
   b. High. Because all of the mass of the product is not measured, then the mass of the limiting reactant will be calculated less. The mass of the starting material is measured but if the mass of the limiting reactant is calculated low, then the mass of the excess reactant is calculated high.

3. The washing of the precipitate removes sodium and chloride ions that may adhere to the dried $Ba_3(PO_4)_2$ precipitate.

*4. Because the mass of $BaSO_4$ is less than the mass of $Ba_3(PO_4)_2$, the yield of $Ba_3(PO_4)_2$ in the experiment is reduced because of the $BaSO_4$ coprecipitation.

LABORATORY QUIZ

1. A mixture of 0.810 g of $BaCl_2 \cdot 2H_2O$ and Na_2SO_4 produces 0.466 g of $BaSO_4$. The limiting reactant in the reaction if Na_2SO_4. Calculate the percent of $BaCl_2 \cdot 2H_2O$ and Na_2SO_4 in the original mixture.   [Answer: 33.5% Na_2SO_4, 66.5% $BaCl_2 \cdot 2H_2O$]

2. A mixture of 0.437 g of $BaCl_2 \cdot 2H_2O$ and 0.284 g Na_2SO_4 is added to water. What is the theoretical yield of $BaSO_4$ precipitate?   [Answer: 0.466 g $BaSO_4$]

3. Describe the test that determines the limiting reactant in a salt mixture that forms a precipitate.   [Answer: See Part B of the Experimental Procedure]

4. The $Ba_3(PO_4)_2$ that passes through the filter is not collected. How does this loss of precipitate affect the reported percentage of the limiting reactant in the salt mixture?
[Answer: See the answer to Laboratory Question No. 1]

A Volumetric Analysis

INTRODUCTION	This basic analytical "wet chemistry" experiment is valuable in the training of chemists. A standardized NaOH solution is prepared in Part A and used in Part B and in Experiments 26–29.
WORK ARRANGEMENT	Individuals. We encourage each student to perform this experiment individually to obtain the hands-on experience of the titration technique.
TIME REQUIREMENT	3 hours. Three trials for Parts A and B are highly recommended. You may decide to perform Part A in the first laboratory period and Part B in a second laboratory period.

LECTURE OUTLINE	1. Follow the Instruction Routine outlined in "To the Laboratory Instructor."
	2. Define and distinguish a stoichiometric point from an endpoint. Also define titrant and analyte.
	3. Present an overview of the experiment and emphasize the importance of the titration technique of volumetric analysis.
	4. Explain why a standard solution of NaOH cannot be made from the dissolution of a measured mass of solid NaOH.
	5. Emphasize that cleanliness and good technique are extremely important for the good results expected in this volumetric analysis. Review Techniques 2, 5, and 16 and the Experimental Procedure very thoroughly with students before the experiment begins.
	6. Inform students (a) if the concentrated NaOH has already been prepared and the $KHC_8H_4O_4$ is dried and (b) if both Parts A *and* B are to be completed.
	7. Review all of Technique 16 carefully with the entire class: the following subject areas should be included in the discussion: • defining an endpoint (and how the indicator has been selected to coincide with the stoichiometric point) • cleaning and preparing a buret with the NaOH titrant • the proper reading of a buret, starting from 0 mL at the top to 50 mL at the bottom (Fig. T.16g). • reading the meniscus in the buret (Fig. T.16a and T.16b) • adding the titrant to the solution (with constant swirling) • stopping titrant addition at the endpoint (the slightest hint of a color change) for phenolphthalein (refer to the color plate).
	8. Advise students that only 2–3 drops of phenolphthalein indicator are needed (if a little is good, a lot is *not* better!!).
	9. The buret should be *thoroughly* rinsed and cleaned at the conclusion of the laboratory period.
	10. Are the students to save their standard NaOH solution for later experiments?

CAUTIONS & DISPOSAL	Review your local procedures for the disposal of chemicals.
	• conc NaOH is as dangerous as conc acids—any spills should be flushed with water and the area covered with $NaHCO_3$.
	• Test solutions and excess reagent chemicals can either be disposed of in a "Waste Acids" container or flushed down the sink, followed by generous amounts of tap water.

- Solid acids from Part B can be discarded in a "Waste Solid Acids" container.

TEACHING HINTS

1. The reproducibility of data in Parts A and B are important. Three trials are a *minimum* for our students.

2. **Part A.1.** To save time, we prepare the conc NaOH and dry the $KHC_8H_4O_4$ for students.

3. **Part A.3.** After the conc NaOH is diluted, it should be well stirred, but not shaken. Shaking a half-filled bottle only reintroduces CO_2 into the solution. The solution should be agitated *only* when the polyethylene bottle is filled. This is a common source of error (producing inconsistent data). Protect the NaOH solution from exposure to the atmosphere at all times to prevent bicarbonate formation:
$$OH^-(aq) + CO_2(aq) \rightarrow HCO_3^-(aq)$$
The *approximate* molar concentration of the NaOH solution is 0.16 *M* NaOH.

4. **Part A.4.** About 0.5 g of $KHC_8H_4O_4$ is required for the reaction of 15 mL of 0.16 *M* NaOH.

5. **Part A.5.** Inspect and approve the student's clean buret before it is filled with NaOH solution. Approval should be based on criteria that no water drops cling to the wall, that the inner wall of the buret is rinsed with titrant (NaOH solution), that the buret is filled with the titrant, and that the air bubbles are removed from the buret tip (see Technique 16.C.1–2).

6. **Parts A.7 and B.** Require students to read the volume of the titrant in the buret with the aid of a blackened line on a white card (Fig. T.16b). The correct titration procedure includes placing a white sheet of paper beneath the receiving flask (Fig. T.16j), using the correct technique for adding the titrant for right-handed (or left-handed, Figs. T.16h and T.16i) chemists, rinsing the wall of the receiving flask with a wash bottle, and adding half-drops of titrant.

7. Explain the importance of stopping titrant addition at the endpoint (Fig. T.16k). Overshooting the endpoint to the same degree in all titrations is incorrect and certainly not an acceptable practice.

8. Students should tightly cap their standardized NaOH solution and save it for Part B and perhaps Experiments 26–29.

CHEMICALS REQUIRED

$KHC_8H_4O_4$ (primary standard, dry in oven at 110°C for 3–4 hours)	2–3 g
phenolphthalein indicator	1 dropper bottle/10 students
NaOH(s)	4 g
or conc NaOH*	5 mL

*For 20 students the conc NaOH solutions should be made up 7 days prior to the laboratory session: slowly dissolve 80 g NaOH in 100 mL water (the reaction is exothermic). Store the NaOH solution in a stoppered polyethylene bottle. Carefully decant the concentrated NaOH solution into a second polyethylene bottle and make it available to the students.[1] The small amount of sediments that accompany the decantate is no problem if the solution is carefully decanted. **Cautions:** *do not store conc NaOH solutions in glass or use a ground glass stoppered bottle. Do not allow conc NaOH to contact the skin.*

SUGGESTED UNKNOWNS

Approximately 75 mL of a strong acid solution is required for an analysis

0.1 *M* HCl
0.15 *M* HCl
0.2 *M* HCl

---

[1]The viscous solution can be vacuum filtered through filter paper of medium porosity. Clean the porcelain filter and flask *immediately* after use.

SPECIAL EQUIPMENT	drying oven, 110°C	1	500-mL reagent bottle, polyethylene	1
	desiccator or desicooler (optional)		ring stand and buret clamp	1
	balance (±0.001 g)		250-mL Erlenmeyer flasks	2
	weighing paper	5	"Waste Acids" container	
	50-mL buret and buret brush	1		

PRELABORATORY ASSIGNMENT

1. A stoichiometric point occurs when stoichiometric amounts of acid and base are present in the system. An endpoint is when the indicator used in the titration changes color.

2. a. Potassium hydrogen phthalate, $KHC_8H_4O_4$
   b. $HC_8H_4O_4^-(aq) + OH^-(aq) \rightarrow C_8H_4O_4^{2-}(aq) + H_2O(l)$

3. $KHC_8H_4O_4$ does not absorb water from the atmosphere.

4. A drop is suspended from the tip of the buret and then either rinsed off with deionized water or touched to the side of the flask and rinsed down the wall with deionized water.

5. a. 0.139 M NaOH
   b. 0.122 g $H_2C_2O_4 \cdot 2H_2O$

6. a. The color change will be from colorless to pink.
   b. Assume that both protons are neutralized at the endpoint; 0.0354 M H_2A

LABORATORY QUESTIONS

1. If the endpoint is consistently surpassed with an excess of NaOH, this will indicate that the NaOH solution is less concentrated than it actually is. As a result, a quantitative determination of the molar concentration of the NaOH in solution will not result.

2. *Too low*. The excess NaOH used indicates that the solution has fewer moles of OH⁻ per volume than actual, resulting in a low molar concentration determination.

3. a. *Too low*. Since the volume dispensed from the buret is the volume of NaOH recorded for neutralization, it implies more NaOH solution (and thus a weaker solution) is required for neutralization.
   b. *Too low*. Since the error in the analysis is a fewer number of moles of NaOH per liter, this means that fewer moles of unknown acid will be reported in the analysis, resulting in a low molar concentration for the acid.

4. *Too low*. The air bubble is recorded as volume of NaOH dispensed. The greater recorded volume of NaOH implies a weaker solution of NaOH and thus its molar concentration will be recorded low.

*5. Phenolphthalein is also an acid and in the titration with NaOH it too is neutralized (when it turns pink). In the analysis, essentially all of the NaOH added should be used for neutralizing the acid being analyzed, *not* the indicator.

*6. The quantity of acid in the flask is fixed by mass—the volume of water added does *not* affect this amount. However, if water is added to the base, this changes the concentration of the base and thus any calculations or interpretations that result.

LABORATORY QUIZ

1. a. Define a stoichiometric point.
   b. Differentiate a stoichiometric point from an endpoint.

2. a. A 0.417-g sample of $KHC_8H_4O_4$ (molar mass = 204.2 g/mol) requires 23.7 mL of a NaOH solution to reach the phenolphthalein endpoint. Calculate the molar concentration of the NaOH solution.   [Answer: 0.0862 M NaOH]
   b. A 16.7-mL volume of this NaOH solution neutralizes 20.0 mL of a hydrochloric acid solution. What is the molar concentration of the HCl solution? [Answer: 0.0719 M HCl]

3. In preparing to standardize a NaOH solution, a student dissolves a $KHC_8H_4O_4$ sample (±0.001 g) in 100 mL of deionized water instead of the recommended 50 mL of water.

Will this affect the measured molar concentration of the NaOH solution? Explain.

[Answer: no]

4. A concentrated NaOH solution is diluted with previously boiled, deionized water to remove traces of CO_2. How do traces of CO_2 affect the concentration of NaOH in solution?   [Answer: CO_2, being acidic, lowers the molar concentration of the NaOH]

5. Drops of titrant cling to the wall of a dirty buret. If a dirty buret is used for standardizing a NaOH solution (the NaOH solution is the titrant) against solid $KHC_8H_4O_4$, is the reported molar concentration of the NaOH high or low?   [Answer: low]

Dry Lab **2A**
Inorganic Nomenclature, I. Oxidation Numbers

INTRODUCTION   The DL2 series of Dry Labs focuses on the naming of inorganic compounds. These Dry Labs are included in this manual because, typically, insufficient time is allotted during lecture.

LECTURE OUTLINE

1. Follow the Instruction Routine outlined in "To the Laboratory Instructor."

2. Review the 8 oxidation number rules. Use a few examples to illustrate the use of the oxidation number rules. Also select examples similar to those presented in the Dry Lab.

3. Use oxidation numbers in explaining the procedure for writing the formulas of compounds.

4. Make an appropriate assignment.

Oxidation numbers

1. a. 2^+ d. 3^- g. 6^+ j. 3^+
    b. 4^+ e. 2^- h. 2^+ k. 1^+
    c. 7^+ f 4^+ i. 3^+ l. 5^+

2. a. 5^+ d. 5^+ g. 4^+ j. 6^+
    b. 3^+ e. 3^+ h. 3^- k. 5^+
    c. 2^+ f. 4^+ i. 5^+ l. 2^+

3. a. 3^+ d. 4^+ g. 3^+ j. 4^+
    b. 5^+ e. 6^+ h. 6^+ k. 6^+
    c. 5^+ f. 2^+ i. 3^+ l. 7^+

4. a. $NaCl$ d. K_3N g. Al_2O_3 j. $CoBr_3$
    b. CaS e. ZnI_2 h. Li_3N k Ag_2O
    c. $FeCl_3$ f. Cr_2O_3 i. Ba_3P_2 l. ZrI_4

Inorganic Nomenclature, II.
Binary Compounds

INTRODUCTION   This second dry lab on inorganic nomenclature focuses exclusively on the naming and writing the formulas of binary compounds, binary acids and hydrates.

LECTURE OUTLINE

1. Follow the Instruction Routine outlined in "To the Laboratory Instructor."

2. Define and review the nomenclature of binary (ionic) salts. For the cations with two common oxidation numbers, review both the "Old" system (the *-ic, -ous* system) and the Stock system of nomenclature. Select some examples similar to those presented in the Dry Lab. Write formulas for binary salts.

3. Suggest which of the cations commonly named by the *-ic, -ous* system should be memorized.

4. Define and review the nomenclature of binary (covalent) compounds of two nonmetals or a metalloid and a nonmetal. Notice that Greek prefixes are used exclusively in their nomenclature and *not* the "*-ic, -ous*" suffixes or the Stock system. Write formulas for these compounds to reinforce the nomenclature rule.

5. Define and review the nomenclature of hydrates and binary acids.

6. Review the procedure for writing formulas.

7. Make appropriate assignments.

Naming Compounds and Writing formulas

1.
   a. sodium phosphide
   b. cesium fluoride
   c. calcium bromide
   d. magnesium oxide
   e. lithium sulfide
   f. ammonium sulfide
   g. calcium hydride
   h. potassium cyanide
   i. strontium hydroxide
   j. calcium nitride
   k. potassium telluride
   l. sodium oxide
   m. calcium carbide
   n. lithium hydroxide
   o. magnesium phosphide

2. a. chromium(II) sulfide
        chromous sulfide
    b. iron(III) oxide
        ferric oxide
    c. chromium(III) iodide hexahydrate
        chromic iodide hexahydrate
    d. copper(I) chloride
        cuprous chloride
    e. lead(II) bromide
        plumbous bromide
    f. mercury(I) chloride
        mercurous chloride
    g. iron(II) sulfide
        ferrous sulfide
    h. iron(III) iodide hexahydrate
        ferric iodide hexahydrate
    i. copper(II) bromide tetrahydrate
        cupric bromide tetrahydrate

    j. cobalt(III) bromide hexahydrate
        cobaltic bromide hexahydrate
    k. tin(II) chloride
        stannous chloride
    l. lead(IV) oxide
        plumbic oxide
    m. tin(IV) fluoride
        stannic fluoride
    n. lead(II) oxide
        plumbous oxide
    o. cobalt(II) oxide
        cobaltous oxide
    p. mercury(II) oxide
        mercuric oxide
    q. copper(I) oxide
        cuprous oxide
    r. mercury(II) sulfide
        mercuric sulfide

3. a. hydrofluoric acid
    b. hydroiodic acid
    c. hydroselenic acid

    d. hydrobromic acid
    e. hydrotelluric acid
    f. hydrochloric acid

4. a. sulfur dioxide
    b. sulfur trioxide
    c. iodine trifluoride
    d. bromine pentafluoride
    e. phosphorus trichloride
    f. sulfur hexafluoride
    g. dinitrogen pentaoxide
    h. dinitrogen tetrasulfide
    i. chlorine trifluoride

    j. silicon tetrachloride
    k. arsenic trifluoride
    l. arsenic pentafluoride
    m. oxygen difluoride
    n. hydrogen chloride
    o. dihydrogen sulfide
    p. xenon tetrafluoride
    q. xenon hexafluoride
    r. selenium tetrachloride

5. a. FeS
    b. $Cu(OH)_2$
    c. $HgCl_2$
    d. CdS
    e. Al_2S_3

    f. $CuCl$
    g. NH_4CN
    h. $MnO \cdot 4H_2O$
    i. MnO_2
    j. $SnCl_2$

    k. CdI_2
    l. TiO_2
    m. $AgCN$
    n. Cr_2O_3
    o. V_2O_5

    p. CaH_2
    q. Fe_2O_3
    r. $CoCl_3 \cdot 6H_2O$
    s. CoO
    t. Hg_2O
    u. $Cu(CN)_2 \cdot 4H_2O$

6. a. $HCl(aq)$
    b. $H_2S(aq)$
    c. $H_2Se(aq)$
    d. $HI(aq)$
    e. NI_3

    f. SiF_4
    g. AsF_5
    h. XeO_4
    i. XeF_6
    j. SO_2

    k. CoI_2
    l. Cl_2O_7
    m. $Au(CN)_3 \cdot 2H_2O$
    n. CrO_3
    o. VF_5

    p. LiH
    q. FeO
    r. $Co_2O_3 \cdot 6H_2O$
    s. CoS
    t. $PbCl_2$
    u. $ZrCl_4$

7. a. $Fe(OH)_3$
    b. Cu_2S
    c. Hg_2Cl_2
    d. $ZnSe$
    e. Na_2S

    f. $Cu(CN)_2$
    g. NH_4I
    h. Mn_2O_7
    i. $MnCl_2$
    j. HgO

    k. $CuCN$
    l. SO_3
    m. I_2O_5
    n. ICl_3
    o. KrF_2

    p. N_2O_4
    q. $AsCl_3$
    r. S_4N_4

Dry Lab 2C
Inorganic Nomenclature, III.
Ternary compounds

INTRODUCTION

Dry Lab 2C is the final dry lab that focuses on the naming of inorganic compounds, focusing on the nomenclature of ternary salts, acid salts, and ternary acids (also called oxoacids).

LECTURE OUTLINE

1. Follow the Instruction Routine outlined in "To the Laboratory Instructor."

2. Define and review the nomenclature of ternary salts. For the cations with two common oxidation numbers, review both the *-ic, -ous* system and the Stock system of nomenclature. Select some examples similar to those presented in the Introduction.

3. Define and review the nomenclature of ternary compounds that may have two or more polyatomic anions, such as SO_3^{2-} and SO_4^{2-}. Practice with the *-ate, -ite* system is important. Wherever necessary, the prefixes *per-* and *hypo-* are used, primarily for the halo-oxyanions, but Table DL2C.2 extends the prefixes and suffixes to less common oxyanions. You should use several examples to illustrate this nomenclature.

4. Define and review the nomenclature of ternary acids. Be sure that you emphasize the relationships between the *-ate salt* to the *-ic acid* and the *-ite salt* to the *-ous acid*.

5. Define and review the nomenclature of acid salts. Again, a few examples will clarify any questions.

6. Make an appropriate assignment.

Naming Compounds and Writing formulas

1.
   a. bromate ion
   b. hypoiodite ion
   c. hypophosphite ion
   d. hyponitrite ion
   e. arsenite ion
   f. bromite ion
   g. iodate ion
   h. sulfite ion
   i. silicate ion
   j. tellurate ion
   k. selenite ion
   l. nitrite ion

2.
   a. sodium sulfate
   b. potassium nitrate
   c. lithium carbonate
   d. calcium phosphate
   e. potassium dichromate
   f. silver(I) chromate
   g. sodium permanganate
   h. potassium acetate
   i. lithium thiosulfate
   j. barium nitrite
   k. silver(I) perchlorate
   l. vanadyl sulfate

3.
   a. iron(II) sulfate heptahydrate
   b. copper(I) cyanide
   c. mercury(II) nitrate monohydrate
   d. copper(II) carbonate
   e. cobalt(II) sulfate heptahydrate
   f. iron(III) hydroxide
   g. chromium(II) cyanide
   h. tin(IV) nitrite
   i. cobalt(III) carbonate
   j. lead(II) acetate trihydrate
   k. iron(III) phosphate
   l. tin(II) nitrate
   m. copper(I) perchlorate
   n. chromium(II) sulfate
   o. chromium(III) phosphate
   p. mercury(I) sulfate

4.
a. hypochlorous acid
b. sulfurous acid
c. phosphoric acid
d. arsenic acid
e. permanganic acid
f. chromic acid
g. boric acid
h. nitrous acid
i. sulfuric acid
j. periodic acid
k. chloric acid
l. bromous acid

5.
a. potassium bicarbonate
   potassium hydrogen carbonate
b. sodium bisulfate monohydrate
   sodium hydrogen sulfate
   monohydrate
c. calcium bicarbonate
   calcium hydrogen carbonate
d. potassium bisulfite
   potassium hydrogen sulfite
e. sodium bisulfide
   sodium hydrogen sulfide
f. sodium bicarbonate
   sodium hydrogen carbonate
g. ammonium dihydrogen phosphate
h. potassium dihydrogen arsenate
i. lithium hydrogen phosphate
j. magnesium hydrogen arsenate
k. ammonium bicarbonate
   ammonium hydrogen carbonate
l. barium bisulfate
   barium hydrogen sulfate

6.
a. $KMnO_4$
b. $CaCO_3$
c. $FeSO_4 \cdot 7H_2O$
d. $Fe_2(C_2O_4)_3$
e. Na_2CrO_4
f. $Ni(NO_3)_2 \cdot 6H_2O$
g. $PbCO_3$
h. $NaNO_2$
i. $Ba(CH_3CO_2)_2$
j. $Ag_2S_2O_3$
k. Na_2SiO_3
l. $Ca(ClO)_2$
m. $KClO_3$
n. $CuIO_3$
o. $(NH_4)_2C_2O_4$
p. $K_2Cr_2O_7$
q. $VO_2(NO_3)_2$
r. $UO_2(CH_3CO_2)_2$
s. Na_3BO_3
t. Na_3AsO_4

7.
a. $H_2SO_4(aq)$
b. $H_3PO_4(aq)$
c. $HNO_3(aq)$
d. $HClO_4(aq)$
e. $HClO_3(aq)$
f. $HOCl(aq)$
g. $HNO_2(aq)$
h. $H_3AsO_4(aq)$
i. $H_2CO_3(aq)$
j. $HBrO_2(aq)$
k. $H_2CrO_4(aq)$
l. $HMnO_4(aq)$
m. $H_2C_2O_4(aq)$
n. $H_2SiO_3(aq)$

8.
a. HF
b. $HBr(aq)$
c. $HBrO(aq)$
d. SO_2
e. P_4O_{10}
f. $HCN(aq)$
g. OsO_4
h. $AuCl_3$
i. NI_3
j. $Cu(OH)_2$
k. Na_2O_2
l. $K_2C_2O_4$
m. $NaHSO_3$
n. $H_2SO_3(aq)$
o. $H_3AsO_3(aq)$
p. $CrCl_3 \cdot 6H_2O$
q. $Al(OH)_3$
r. $Co(NO_3)_2 \cdot 6H_2O$
s. $AgCN$
t. $H_3BO_3(aq)$
u. S_2O_7
v. $H_2S_2O_3(aq)$
w. SiO_2
x. $HCH_3CO_2(aq)$
y. $VOCl_2$
z. $CaHPO_4$

9.
a. $Ba(CH_3CO_2)_2 \cdot 2H_2O$
b. $CuCl$
c. CdI_2
d. SnO_2
e. VF_5
f. IF_3
g. SiF_4
h. $(NH_4)_2S$
i. FeC_2O_4
j. HgO
k. $LiClO$
l. CaH_2
m. Cu_2O
n. Na_2SiO_3
o. N_2S_4
p. V_2O_5
q. $TiCl_4$
r. $Sc(NO_3)_3$
s. $FePO_4 \cdot 6H_2O$
t. $Pb(CH_3CO_2)_2$
u. Ca_3N_2
v. $Ni(CH_3CO_2)_2 \cdot 6H_2O$
w. $Cr(CH_3CO_2)_2$
x. $Hg_2(NO_3)_2$
y. $Fe_2(CrO_4)_3$
z. $(NH_4)_2Cr_2O_7$

Experiment 10
Periodic Table and Periodic Law

INTRODUCTION

This experiment attempts to show the periodic trends of the chemical and physical properties of the elements. The trends of the physical properties of elements are studied from data plots of various physical measurements versus atomic number. Trends in the chemical properties of the elements are determined from comparative observations of chemical reactions of elements positioned adjacent to each other in the Periodic Table.

WORK ARRANGEMENT

Individuals for Parts A and B; Partners for Parts C, D and E.

TIME REQUIREMENT

1.5 hours for Parts A and B; 2.5 hours for Parts C, D and E. Because of the length of the experiment, you may choose to delete part(s) of the Experimental Procedure.

LECTURE OUTLINE

1. Follow the Instruction Routine outlined in "To the Laboratory Instructor."

2. Review the various sections of the periodic table and the definitions of the properties plotted in Parts A and B. Emphasize that an understanding of the periodic trends of physical and chemical properties are important for predictions of chemical phenomena.

3. Data points for the periodic trends in Parts A and B are to be connected by straight (and dotted, where data are missing) lines. This graphing procedure is slightly different from that described in Appendix C.

4. This is a long experiment (3–4 hours). Be organized!!! You may choose to omit some sections of Parts C, D, E or the Laboratory Questions. On occasion we assign Parts A and B as a take home assignment.

5. Divide the class into two groups and proceed through the experiment accordingly:
   Group I.   A, B, C, D, E
   Group II.   C, D, E, A, B

CAUTIONS & DISPOSAL

- Care should be exercised in handling 3 *M* and 6 *M* HCl, 6 *M* HNO$_3$, 6 *M* NaOH and conc H$_2$SO$_4$.

- Advise students to dispose of the test solutions for **Parts C and D** in the "Waste Halogens Container." The test solutions from **Part E** can be discarded in the sink, followed by copious amounts of tap water.

- The demonstration of the reaction for sodium and water (**Part E.1**) should be performed using your *best* laboratory safety technique.

TEACHING HINTS

1. **Parts A and B.** Notice the instructions for completing the graphs in the Experimental Procedure. Also review Appendix C. Advise students to properly label the axes and title the graph.

2. **Part C.1.** These samples are for viewing only.

3. **Part C.2.** Chlorine is toxic. Observe closely the procedure by which students generate chlorine. The chlorine, being nonpolar, is soluble in the mineral oil, the upper layer.
   $$ClO^-(aq) + HCl(aq) + H^+(aq) \rightarrow Cl_2(mineral\ oil) + H_2O(l)$$

4. **Parts C.3, 4.** The conc H$_2$SO$_4$ oxidizes the bromide to bromine and the iodide to iodine, both of which are soluble in the mineral oil.

5. **Part D.1.** The formation of Cl$_2$ from ClO$^-$ is favored in an HCl(*aq*) solution. Therefore, the addition of several drops of 6 *M* HCl (see Teaching Hint 3 above) promotes the

reaction rates

$$Cl_2(mineral\ oil) + 2\ Br^-(aq) \rightarrow 2\ Cl^-(aq) + Br_2(mineral\ oil)$$
and $Cl_2(mineral\ oil) + 2\ I^-(aq) \rightarrow 2\ Cl^-(aq) + I_2(mineral\ oil)$

6. **Part D.2.** $Br_2(mineral\ oil) + 2\ I^-(aq) \rightarrow 2\ Br^-(aq) + I_2(mineral\ oil)$
   but $Br_2(mineral\ oil) + 2\ Cl^-(aq) \rightarrow$ no reaction

7. **Part D.3.** No reactions should be observed; only the violet I_2 should appear in the mineral oil layer.

8. **Part D.4.** a. CaF_2 has a low solubility; all other Ca^{2+} halide salts are soluble.
   b. AgF is soluble; all other Ag^+ halide salts are insoluble—white (Cl^-) to yellow (I^-).
   $AgCl$ forms $[Ag(NH_3)_2]^+$ with ammonia.
   c. $[FeF_6]^{3-}$ forms; no other evidence of change occurs.

9. **Part E.1.** The reaction of sodium with water is safe *if* the directions are closely followed; however, the reaction must be performed by you or a more confident instructor–*not* a student. If you do *not* feel comfortable performing the experiment, *don't*!
   $$2\ Na(s) + 2\ H_2O(l) \rightarrow 2\ NaOH(aq) + H_2(g)$$

10. **Part E.2.** Mg reacts more rapidly with HCl than does Al to produce H_2 gas. $Al(OH)_3$ is amphoteric, $Mg(OH)_2$ is not. Therefore $Al(OH)_3$ dissolves in excess OH^- forming $Al(OH)_4^-$.

    $Al^{3+}(aq) + 3\ OH^-(aq) \rightarrow Al(OH)_3(s)$ $\quad Mg^{2+}(aq) + 2\ OH^-(aq) \rightarrow Mg(OH)_2(s)$
    $Al(OH)_3(s) + OH^-(aq) \rightarrow Al(OH)_4^-(aq)$ $\quad Mg(OH)_2(s) + OH^-(aq) \rightarrow$ no reaction

11. **Part E.3.** $Mg(OH)_2$ is least soluble of the Group IIA hydroxides.

12. **Part E.4.** $Na_2SO_3(aq) + 2\ HCl(aq) \rightarrow 2\ NaCl(aq) + H_2O(l) + SO_2(g)$. $SO_2(g)$ is an acidic oxide and therefore turns blue litmus red. Sulfuric acid is a nonvolatile acid and therefore its vapors do not affect litmus, but, of course, the sulfuric acid solution turns blue litmus red.

CHEMICALS REQUIRED

Set up on the display table:
sodium (under cyclohexane)
magnesium in a beaker
aluminum in a beaker
silicon in a beaker
sulfur in a beaker

5% NaClO (commercial laundry bleach)		3 M NH$_3$ (dropper bottle)	3 mL
	3 mL	6 M HNO$_3$ (dropper bottle)	0.5 mL
mineral oil	1 mL	0.1 M Fe(NO$_3$)$_3$ (dropper bottle)	3 mL
6 M HCl (dropper bottle)	2 mL	Na metal in Al foil	pea-size
6 M KBr (dropper bottle)	2 mL	phenolphthalein (dropper bottle)	0.5 mL
conc H$_2$SO$_4$ (dropper bottle)	2 mL	Mg strip	1 cm
6 M KI (dropper bottle)	1 mL	Al strip	1 cm
KBr(s)	0.5 g	3 M HCl (dropper bottle)	2 mL
KI(s)	0.5 g	6 M NaOH (dropper bottle)	2 mL
NaCl(s)	0.5 g	0.10 M MgCl$_2$ (dropper bottle)	1 mL
NaF(s)	0.1 g	0.10 M CaCl$_2$ (dropper bottle)	1 mL
2 M Ca(NO$_3$)$_2$ (dropper bottle)	3 mL	0.01 M NaOH (dropper bottle)	1 mL
0.1 M AgNO$_3$ (dropper bottle)	3 mL	Na$_2$SO$_3$(s)	0.1 g
		0.1 M H$_2$SO$_4$ (dropper bottle)	1 mL

SPECIAL EQUIPMENT	"Waste Halogens" container steel wool litmus paper (red/blue) 1 vial of each 24-well plate and Beral pipets	200-mm test tube safety shield aluminum foil Bunsen burner

PRELABORATORY ASSIGNMENT

3. a. chlorine, Cl_2
   b. bromine, Br_2
   c. iodine, I_2

4. a. less
   b. more

5. a. H.G.J. Moseley
   b. Dmitri Mendeleev
   c. John Newlands
   d. Lothar Meyer
   e. Dmitri Mendeleev
   f. Johann W. Döbereiner

LABORATORY QUESTIONS

1. a. increase   f. fluorine
   b. decrease   g. increases
   c. decreases   h. decrease
   d. IVA   i. decrease
   e. 0, noble gases

2. number

3. In Period 3, sodium is a soft metal, aluminum is a harder metal, and argon, at the far right of the period, is a gas. Fluorine and chlorine are gases, bromine is a liquid, and iodine is a solid.

4. Of the halogens studied in this experiment, Cl_2 is most reactive, I_2 is least reactive; Cl_2 reacts with I^-, but I_2 does not react with Cl^-.

5. The reactivity of the metals decreases across the period of representative elements. Sodium metal reacts vigorously with water and the oxygen of air, magnesium and aluminum metals are safe to handle with the hands.

LABORATORY QUIZ

1. _____ is the scientist that is credited with establishing the current format for the periodic chart.   [Answer: Mendeleev]

2. a. The _____ is the number of protons in the nucleus.
                                                                    [Answer: atomic number]
   b. The Groups IIA elements are also called the _____ metals .
                                                                    [Answer: alkaline-earth]
   c. The halogens are the Group _____ elements.   [Answer: VIIA]
   d. The representative elements are designated as the Group _____ elements.
                                                                    [Answer: A]
   e. The symbols of at least two noble metals are _____ and _____.
                                                        [Answer: see lab manual, p. 130]

3. List the formula, color, and physical state (at room temperature and pressure) of chlorine, bromine and iodine.   [Answer: $Cl_2(g)$, yellow-green, $Br_2(l)$, red-brown; $I_2(s)$, violet]

4. Circle the best choice: (Answers are boxed)

a. the most soluble hydroxide $Ca(OH)_2$ $Sr(OH)_2$ ☐$Ba(OH)_2$

b. the highest ionization energy   ☐Cl   Br   I

c. the highest electronegativity   ☐O   S   P

d. the smallest atomic radius   ☐Mg   Na   K

e. the most reactive Br_2 I_2 ☐Cl_2

f. an alkali metal   ☐Li   Mg   Al

g. a halogen   ☐I   O   Na

h. a noble gas   Au F_2 ☐Ar

i. an alkaline-earth metal   Cs   ☐Ba   Ho

j. a rare-earth metal   ☐Ce   Cs   Hg

k. a transition metal   Pb   ☐Fe   Al

l. a metalloid   Sn   ☐As   Pb

m. an inner transition metal   W   ☐Nd   Cd

INTRODUCTION

A hands-on understanding of electronic energy states is difficult to portray to general chemistry students in the laboratory. An appropriate setup for viewing spectra is quite elaborate and, therefore, is limited to the number of students who can use it. In this experiment, students first observe the colors that result from the flame tests of various salts. Thereafter, the hydrogen spectrum is studied using the color plate insert in the manual; the wavelengths of the various emission lines are determined and then the quantification of energy is realized through the Bohr calculations. A value of the Rydberg constant is determined. Students then identify an element based on its emission spectrum appearing on the color plate. The wavelengths of the emission spectrum on the color plate are determined and compared to experimental wavelengths reported in the literature. A "match" identifies the element.

WORK ARRANGEMENT

Partners (as a team) for Parts A and B; individuals for Part C.

TIME REQUIREMENT

2 hours

LECTURE OUTLINE

1. Follow the Instruction Routine outlined in "To the Laboratory Instructor."

2. **Part A.** The platinum or nichrome wire must be thoroughly cleaned before each test. The flame tests for each element should be repeated for clarity, confirmation, and confidence. On occasion we assign an unknown salt for identification. A few flame tests are shown on the color plate in the manual.

3. **Part B.1.** Spectrum 7 on the color plate shows the hydrogen spectrum; the mercury spectrum (bottom) and the visible continuous spectrum (top) are also shown. From the known wavelengths of the mercury spectrum, Table 11.1, students estimate the emission wavelengths of the hydrogen atom.

4. **Part B.2.** From the estimated wavelengths, students calculate $\Delta E_{atom} = E_{photon}$ (Eq. 11.1) and $1/\lambda$. Using the *approximate* value of the Rydberg constant, $R = 1.1 \times 10^{-2}$ nm^{-1}, n_h for each line in the H-spectrum is calculated (rearrangement of Eq. 11.2). The value of n_h is then be rounded off to the nearest whole number before calculating $1/n_h^2$. The slope of a plot of $1/\lambda$ versus $1/n_h^2$ determines a more accurate value of the Rydberg constant.

5. **Part C.** Students are assigned one of spectrum (1–10) from the color plate and, from a calibration of the spectrum with the known emission lines of the mercury spectrum, record the principle emission lines of the unknown element. The estimated emission lines of the spectrum are compared with those listed in Table 11.2 for its identification.

CAUTIONS & DISPOSAL

- Conc HCl is used in Part A. All spills should be cleaned up immediately.

TEACHING HINTS

The spectral key is

Spectrum No.	1	2	3	4	5	6	7	8	9	10
Elements	He	Ne	Na	Zn	Cd	Cs	H	Rb	Tl	K

1. An unknown salt can be assigned for Part A. Students tend to make closer observations of the flame tests (also used in the qualitative analysis experiments) if an unknown is assigned.

2. A suggested procedure for calibrating the color plate: the measured distance between the 546.0 nm and 435.8 nm lines for the mercury spectrum on the color plate is 40 mm, making each millimeter on the color plate 2.75 nm. Marking 550 nm as a reference point on the color plate, the line spectra of the other elements agree very closely with the reported lines of the spectra in Table 11.2.

3. **Corrections (first printing only):** The line spectrum for neon (Spectrum 2 on the color plate) should be shifted *left* by 6 mm; a wavelength in the cadmium spectrum (Table 11.2) should be changed from 518.6 nm to 508.6 nm.

4. Assistance with the calculations in Part B.3 is necessary for many students.

5. Calculated values for n_{fi} are to be rounded to whole numbers.

6. The *exact* data for the hydrogen spectrum are

λ(nm)	color	ΔE_{atom}(J/photon)	$1/\lambda$(nm^{-1})	n_{fi}	n_{fi}*	$1/n_{fi}^2$
410.2	ultraviolet	4.85 x 10^{-19}	0.00244	5.96	6	0.0278
434.0	violet	4.58 x 10^{-19}	0.00230	4.95	5	0.0400
486.1	blue	4.09 x 10^{-19}	0.00206	3.99	4	0.0625
656.3	red	3.03 x 10^{-19}	0.00152	2.99	3	0.1111

*rounded off to the nearest whole number

Data For Determining Rydberg Graph

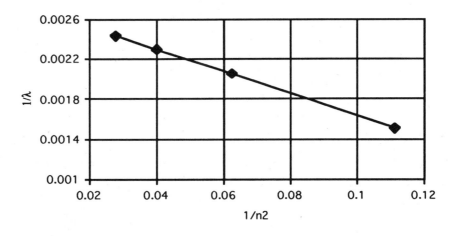

7. Approve the $1/\lambda$ versus $1/n_{fi}^2$ graph in Part B.

8. The accepted value for R = 1.09678 x 10^{-2} nm^{-1}

CHEMICALS REQUIRED				
conc HCl (dropper bottle)	5 mL	BaCl$_2$ green flame test	few crystals
CaCl$_2$ brick-red flame test	few crystals	LiCl red flame test	few crystals
CuCl$_2$ green flame test	few crystals	SrCl$_2$ crimson flame test	few crystals
NaCl orange-yellow flame test	few crystals	KCl violet flash through cobalt glass plate	few crystals

SPECIAL EQUIPMENT	
flame test wire (platinum or nichrome)	1
watchglass	2
cobalt glass plate	1
spectroscope (optional)	1
Bunsen burner	

1. For an absorption spectrum electrons absorb quantized amounts of energy and are excited, the remaining energy is transmitted. As a result, black absorption lines appear in a continuous spectrum. In an emission spectrum, electrons emit quantized amounts of energy when they de-excite and only these lines of EM radiation appear on a black background.

2. a. 3.02×10^{-19} J
   b. red

3. a. 121.2 nm, using $R = 1.1 \times 10^{-2}$ nm^{-1}
   b. ultraviolet
   c. 1.64×10^{-18} J

4. An atom is an "excited state" when an electron has absorbed energy and has been promoted to a higher energy state.

5. $n_f = 2$ for the visible spectrum of hydrogen.

6. 6 lines; two lines in the visible spectrum:
   $n_h = 3 \rightarrow n_f = 2$ and $n_h = 4 \rightarrow n_f = 2$ electron transitions.
   The other four lines are:
   $n_h = 4 \rightarrow n_f = 1, n_h = 3 \rightarrow n_f = 1, n_h = 2 \rightarrow n_f = 1, n_h = 4 \rightarrow n_f = 3$

**LABORATORY
QUESTIONS**

1. a. Infrared region of the EM spectrum
   b. 1094 nm

2. Each ion has its own set of electron energy levels. The de-excitation of an electron emits a photon with a characteristic color, corresponding to the unique energy difference for that electron in that particular ion.

3. Yellow-orange light emission is in the 550-600 nm region of the EM spectrum.

4. Blue light emission is in the 400-450 nm region of the EM spectrum.

5. a. Excited sodium ions produce the orange-yellow color.
   b. Excited strontium and/or lithium ions produce the red color in starbursts.

LABORATORY QUIZ

1. The Rydberg constant and corresponding equations are applicable to only one element. Which one?

2. Explain why the line spectra of different elements are different.

3. What is the source of the appearance of a line in a line spectrum?

4. a. Calculate the energy of a 500-nm photon. $h = 6.63 \times 10^{-34}$ J•s/photon
   [Answer: 3.98×10^{-19} J]
   b. Is this photon in the visible region of the electromagnetic spectrum?
   [Answer: Yes, green region]
   c. If one mole of electrons are pulsed, emitting 500-nm radiation, how much energy is emitted? [Answer: 2.40×10^5 J]
   d. How many liters of water can be heated from 25°C to 100°C using energy in Part 1c? The specific heat of water is 4.18 J/(g•°C). [Answer: 0.765 L]

5. What color is the flame test for:
   a. Ca^{2+} [Answer: brick-red]
   b. Cu^{2+} [Answer: green]
   c. Ba^{2+} [Answer: green]
   d. Sr^{2+} [Answer: crimson]
   e. Na^+ [Answer: orange-yellow]

6. An intense line in the indium emission spectrum occurs at 451.1 nm. Calculate the energy of a photon that results from this electron transition. [Answer: 4.40×10^{-19} J]

Molecular Geometry

INTRODUCTION

Molecular models are used to construct a selection of molecules and molecular ions. Valence bond (VB) and valence shell electron pair (VSEPR) bonding theories are used to account for the bonding and the geometry of molecules and molecular ions.

Students who work in teams (or groups) learn about VB and VSEPR theories—the similarities, differences, and how they complement each other. In addition, discussions of bond angles, geometric shapes, 3-D molecular shapes, and polarity become more meaningful.

WORK ARRANGEMENT

Partners (or groups, depending upon the number of sets of molecular models).

TIME REQUIREMENT

2.5 hours

LECTURE OUTLINE

1. Follow the Instruction routine outlined in "To the Laboratory Instructor."

2. Review Lewis symbols and the Lewis formulas for covalent systems. A review of VB and VSEPR theories may require 20–30 minutes. Use Figures DL3.2–DL3.4 in the manual and Table DL3.1 in your discussion.

3. Identify the 5 basic orientations of electron-pairs or hybrid orbitals for molecules and molecular ions (Table DL3.1) as predicted by VB and VSEPR theories. Distinguish between the geometric shape of the electron-pairs and hybrid orbitals in a molecule versus the 3-D shape of a molecule.

4. A short discussion of polar bonds and polar molecules (Figures DL3.5–DL3.7) saves some laboratory time.

5. Make your assignments. We assign about 10 molecules or molecular ions to each student group; they build the molecular model for each and determine the bonding and nonbonding orbitals (or electrons pairs), the hybridization, the VSEPR formula, and the geometry of the molecule or molecular ion. The polarity of the molecules/molecular ions is predicted. An unknown is assigned to each individual, generally from those listed in the Procedure of the Report Sheet.

SPECIAL EQUIPMENT

Molecular model set   1/2 to 3 students

For cheap, molecular model sets (about $3 at 1994 prices), we have placed in a zip-lock bag: five $1\,^1/_2$-in. styrofoam balls (each painted a different color), six 1-in. styrofoam balls (no color change), and ten tooth picks. This set is actually *better* than a commercial set of molecular models because students must establish the correct geometry, including bond angles, for each molecule and molecular ion.

TEACHING HINTS

1. Mingle among the student groups to ask questions about Lewis formulas, VB theory, VSEPR theory, polarity, 3-D shape, etc.

2. Formal charge is *not* considered in the construction and description of the molecules and molecular ions.

3. **Corrections (first printing only)** in Table DL3.1: ICl^{2-} should be ICl_2^- and ICl^{4-} should be ICl_4^-.

Molecule or Molecular Ion	Bonding Orbitals or Pairs	Non-bonding Orbitals or Pairs	Hybrid-ization	VSEPR Formula	3-D Shape	Polar or Nonpolar
2. CF_3Cl	4	0	sp^3	AX_4	tetrahedral	P
3. H_2O	2	2	sp^3	AX_2E_2	V-shaped	P
4. H_2S	2	2	sp^3	AX_2E_2	V-shaped	P
5. NH_3	3	1	sp^3	AX_3E	trigonal pyramid	P
6. NH_4^+	4	0	sp^3	AX_4	tetrahedral	NP
7. AsF_3	3	1	sp^3	AX_3E	trigonal pyramid	P
8. BF_4^-	4	0	sp^3	AX_4	tetrahedral	NP
9. H_3O^+	3	1	sp^3	AX_3E	trigonal pyramid	P
10. PO_4^{3-}	4	0	sp^3	AX_4	tetrahedral	NP
11. ClO_2^-	2	2	sp^3	AX_2E_2	V-shaped	P
12. GaI_3	3	0	sp^2	AX_3	trigonal planar	NP
13. PCl_2F_3	5	0	sp^3d	AX_5	trigonal bipyramid	P or NP
14. BrF_3	3	2	sp^3d	AX_3E_2	T-shaped	P
15. SF_4	4	1	sp^3d	AX_4E	unsym. tetrahedral	P
16. XeF_2	2	3	sp^3d	AX_2E_3	linear	NP
17. PF_5	5	0	sp^3d	AX_5	trigonal bipyramid	NP
18. AsF_5	5	0	sp^3d	AX_5	trigonal bipyramid	NP
19. SF_6	6	0	sp^3d^2	AX_6	octahedral	NP
20. SiF_6^{2-}	6	0	sp^3d^2	AX_6	octahedral	NP
21. IF_4^+	4	1	sp^3d	AX_4E	unsym. tetrahedral	P
22, IF_4^-	4	2	sp^3d^2	AX_4E_2	square planar	NP
23. XeF_4	4	2	sp^3d^2	AX_4E_2	square planar	NP
24. CH_3^-	3	1	sp^3	AX_3E	trigonal pyramid	P
25. CH_3^+	3	0	sp^2	AX_3	trigonal planar	NP
26. PF_3	3	1	sp^3	AX_3E	trigonal pyramid	P
27. PF_4^+	4	0	sp^3	AX_4	tetrahedral	NP
28. SbF_6^-	6	0	sp^3d^2	AX_6	octahedral	NP
29. SnF_4	4	0	sp^3	AX_4	tetrahedral	NP
30. SnF_2	2	1	sp^2	AX_2E	V-shaped	P
31. SiH_4	4	0	sp^3	AX_4	tetrahedral	NP
32. SnF_6^{2-}	6	0	sp^3d^2	AX_6	octahedral	NP
33. SO_4^{2-}	4	0	sp^3	AX_4	tetrahedral	NP

PRELABORATORY ASSIGNMENT

1. a. The bond in valence bond theory is the overlap of two valence shell atomic orbitals on adjacent atoms.
    b. The bond in VSEPR theory is the electron pair.

2.

a. $: S : Cl :$ $: Cl :$

b. $: Cl :$ $: Cl : C : Cl :$ $: Cl :$

c. $: F : N : F :$ $: F :$

3. a. trigonal bipyramid g. 3 m. 2
   b. linear h. 6 n. none
   c. 4 i. trigonal planar o. 1
   d. 3 j. octahedral
   e. 4 k. 6
   f. 3 l. 7

4. a. 3 f. three sp^2 orbitals l. V-shaped
   b. 2 g. 1 m. 4
   c. unsymmetrical h. 2 n. V-shaped
      tetrahedron i. 1
   d. trigonal bipyramid j. 2
   e. sp^3d k. 1

5. a. polar e. yes, polar bonds
   b. polar f. non-polar
   c. polar
   d. non-polar

LABORATORY QUESTIONS

1. a. SO_2 has sp^2 hybridization of atomic orbitals on the sulfur atom. CO_2 has sp hybridization of atomic orbitals on the carbon atom.

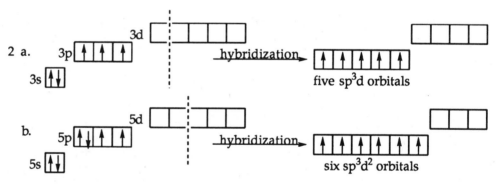

   b. SO_2 should have a V-shaped structure; CO_2 should have a linear structure.

2 a. 3p, 3s → hybridization → five sp^3d orbitals (3d)

   b. 5p, 5s → hybridization → six sp^3d^2 orbitals (5d)

3. $TeF_2(CH_3)_4$ should have an octahedral structure.

4. The bond angle for OF_2 should be < 109.5°.

5. XeO_3 should have a trigonal pyramidal structure with a bond angle of <109.5°.

Inorganic Compounds and Metathesis Reactions

INTRODUCTION

A general exposure to the physical and chemical properties of inorganic compounds is the purpose of this experiment. Our experience has shown that general chemistry students are ignorant of many facts about chemicals that chemists and instructors assume as common knowledge. For example, the facts that silver chloride is a white, crystalline solid, nickel salts are generally green, copper salts are blue, and permanganate ions are purple are informative to students and help close the communication gap between students and instructors and between theory and reality.

WORK ARRANGEMENT

Partners.

TIME REQUIREMENT

2.5 hours

LECTURE OUTLINE

1. Follow the Instruction Routine outlined in "To the Laboratory Instructor."

2. Divide the class into two groups; one to start on Part A and the other on Part B.

3. Review the seven-step procedure for writing a net ionic equation *as the experiment is being performed* and as the reaction is observed.

CAUTIONS & DISPOSAL

- **Part B.** In general, the chemicals used in this experiment present no real danger—only carelessness will cause a chemical accident.

- **Part B.2c.** Detect the odor of the ammonia according to Technique 17A.

- Dispose of the test solutions in the "Waste Liquids" container at the conclusion of Part B.

TEACHING HINTS

1. **Part A.** The samples for Part A are selected from the stockroom shelf, almost at random, and placed in 150-mm test tubes. Nickel, copper, ferrous, ferric, cobalt, permanganate, dichromate, and carbonate salts are selected for the display. White salts are also selected so that students do perceive that all salts have color. A number of watch glasses are made available in the event that a student wants to remove a few crystals (or drops) from the test tube for a closer examination. We use about 50 different chemicals for the display.

2. **Part B.** Chemicals in a 24-well plate are systematically arranged to minimize waste and contamination. The major observational error arises from using contaminated Beral pipets.

   The twelve (12) inorganic compounds listed in Part B are only suggestions. Others may be added or substituted at the discretion of the instructor. Each pair of test solutions is studied according the seven-step analysis described in the Introduction. Observation of the reactions is critical. Partners openly discuss each test.

3. Students may require some assistance in writing the equations for each of the seven steps on the Report Sheet. It is important for the laboratory instructor to explain the differences between a molecular, ionic (Step 6), and net ionic equation (Step 7).

CHEMICALS REQUIRED

Part A. Prepare in 150-mm test tubes approximately 50 different chemicals with characteristic colors and physical states. Display them in test tube racks. One gram samples are sufficient. Include a number of nickel, copper, ferrous, ferric, cobalt, permanganate, dichromate, and carbonate salts.

Part B.

$CaCO_3$	0.1 g	NH_4Cl	0.1 g
3.0 M HCl	3 mL	Na_2CO_3	0.1 g
3.0 M NaOH	3 mL	$NiCl_2 \cdot 6H_2O$	0.1 g
universal indicator (dropper bottle)	0.1 mL	$Na_3PO_4 \cdot 12H_2O$	0.1 g
$FeCl_3 \cdot 6H_2O$	0.1 g*	$CuSO_4 \cdot 5H_2O$	0.1 g
$CoCl_2 \cdot 6H_2O$	0.1 g	$BaCl_2 \cdot 2H_2O$	0.1 g
$AgNO_3$	0.1 g		

*Solutions of the soluble salts can be substituted for the solids. Prepare 2 mL of a 0.1 mol/L solution of each salt for each student group.

SPECIAL EQUIPMENT	24-well plate and Beral pipets
	watchglasses 1/student
	universal indicator color chart

SUGGESTED UNKNOWNS	We issue unknowns from the list in Part B.1; if students use solids in Part B, we issue a 0.1 mol/L unknown salt solution; if students use 0.1 mol/L solutions in Part B, we issue a solid.

REPORT SHEET INFORMATION	**Part A. Identification**

1. Transition metal cations tend to have a color. Displayed examples may include the normally blue color of copper(II) salts, green color of nickel(II) salts, and the orange-brown colors of iron(III) salts.

2. Most anions are colorless (white). The carbonate, phosphate, sulfate, chloride, etc. anions contribute no color. Anions that contain a transition metal ion generally have a color, such as the permanganate and dichromate ions.

3. In general, ionic compounds tend to be solids and covalent (or molecular) compounds tend to have low melting points or are liquids at room temperature. Displayed examples may include water, ethanol, hexane, . . .

4. Hydrated salts tend to be crystalline solids with large facets that are highly reflective. The cations are generally transition metal cations. Displayed examples may include $CuSO_4 \cdot 5H_2O$ or $CoSO_4 \cdot 6H_2O$

Part B. Metathesis Reactions

Evidence of reaction and the net ionic equation for the other test reactions:

a. gas evolved $2\,H^+(aq) + CO_3{}^{2-}(aq) \rightarrow H_2O(l) + CO_2(g)$

b. heat evolved $H^+(aq) + OH^-(aq) \rightarrow H_2O(l) + \text{heat}$

c. odor of ammonia $NH_4{}^+(aq) + OH^-(aq) \rightarrow NH_3(g) + H_2O(l)$

d. red-brown precipitate $Fe^{3+}(aq) + 3\,OH^-(aq) \rightarrow Fe(OH)_3(s)$

e. no reaction

f. white precipitate $Ag^+(aq) + Cl^-(aq) \rightarrow AgCl(s)$

g. white precipitate $2\,Ag^+(aq) + CO_3{}^{2-}(aq) \rightarrow Ag_2CO_3(s)$

h. blue precipitate $Cu^{2+}(aq) + CO_3{}^{2-}(aq) \rightarrow CuCO_3(s)$

i. white precipitate $Ba^{2+}(aq) + SO_4{}^{2-}(aq) \rightarrow BaSO_4(s)$

j. blue precipitate $3\,Cu^{2+}(aq) + 2\,PO_4{}^{3-}(aq) \rightarrow Cu_3(PO_4)_2(s)$

k. green precipitate $3\,Ni^{2+}(aq) + 2\,PO_4{}^{3-}(aq) \rightarrow Ni_3(PO_4)_2(s)$

Generalizations

	SO_4^{2-}	CO_3^{2-}	OH^-	PO_4^{3-}	NO_3^-	Cl^-
Cations Forming Insoluble Salts	Ba^{2+}	Cu^{2+}, Ca^{2+}, Ag^+	Fe^{3+}	Cu^{2+}, Ni^{2+}		Ag^+

	SO_4^{2-}	CO_3^{2-}	OH^-	PO_4^{3-}	NO_3^-	Cl^-
Cations Forming Soluble Salts	Na^+, Cu^{2+}	Na^+	Na^+, H^+	Na^+	Na^+, Co^{2+}, Ag^+	Na^+, Cu^{2+}, Co^{2+}, Fe^{3+}, H^+, Ca^{2+}, Ni^{2+}, Ba^{2+}, NH_4^+

PRELABORATORY ASSIGNMENT

1. a. formation and color of a precipitate
    b. evolution of a gas
    c. evolution or absorption of energy (heat)
    d. change in color
    e. change in acidity of basicity of the solution

2. a. $3\,AgNO_3(aq) + AlCl_3(aq) \rightarrow 3\,AgCl(s) + Al(NO_3)_3(aq)$
    b. $CuSO_4(aq) + Pb(NO_3)_2(aq) \rightarrow Cu(NO_3)_2(aq) + PbSO_4(s)$
    c. $K_2CO_3(aq) + CdI_2(aq) \rightarrow CdCO_3(s) + 2\,KI(aq)$
    d. $3\,FeSO_4(aq) + 2\,K_3PO_4(aq) \rightarrow Fe_3(PO_4)_2(s) + 3\,K_2SO_4(aq)$
    e. $CaCO_3(s) + 2\,HNO_3(aq) \rightarrow Ca(NO_3)_2(aq) + H_2O(l) + CO_2(g)$

3. ionic equation: $Cu^{2+}(aq) + SO_4^{2-}(aq) + Pb^{2+}(aq) + 2\,NO_3^-(aq) \rightarrow$
    $$Cu^{2+}(aq) + 2\,NO_3^-(aq) + PbSO_4(s)$$
    net ionic equation: $Pb^{2+}(aq) + SO_4^{2-}(aq) \rightarrow PbSO_4(s)$

LABORATORY QUESTIONS

1. An acid and a CO_3^{2-} salt generate $CO_2(g)$.

2. AgBr and AgI are predictably insoluble salts because AgCl is also insoluble (see Appendix G).

3. Most salts of Na^+ and NO_3^- are soluble salts.

4. Most Cl^- salts and SO_4^{2-} salts are soluble, whereas most CO_3^{2-} salts are insoluble.

LABORATORY QUIZ

1. List four observations that identify the occurrence of a chemical reaction.

2. a. copper salts generally have a _____ color.
    b. nickel salts generally have a _____ color.
    c. sodium salts generally have a _____ color.
    d. cobalt salts generally have a _____ color.

3. For a mixture of Na_2CO_3 and $CuSO_4$ salts in an aqueous solution:
    a. identify the evidence of a chemical reaction:
    b. write the molecular equation:
    c. write the ionic equation:
    d. write the net ionic equation:

4. For a mixture of Na_3PO_4 and $CoCl_2$ salts in an aqueous solution:
    a. identify the evidence of a chemical reaction:
    b. write the molecular equation:
    c. write the ionic equation:
    d. write the net ionic equation:

5. Circle the *insoluble* salts: $CuSO_4$, $Co(NO_3)_2$, $CaCO_3$, $Ni_3(PO_4)_2$, $BaCl_2$, $BaSO_4$, $CoCl_2$, $Fe(OH)_3$ [Answer: $CaCO_3$, $Ni_3(PO_4)_2$, $BaSO_4$, $Fe(OH)_3$]

Acids and Bases; pH

INTRODUCTION	The chemical properties of acids and bases are often discussed in general chemistry lecture, but never fully realized or systematically studied. This experiment distinguishes between acid strengths, both as acids and as oxidizing agents, and base strengths. Simple interpretations of the acidity of solutions is expressed in terms of pH. Our students thoroughly enjoy the qualitative observations in this experiment.
WORK ARRANGEMENT	Partners. The exchange of opinions and observations is beneficial in this experiment.
TIME REQUIREMENT	3 hours. Because of its length, you may wish to select certain parts of the Experimental Procedure.

LECTURE OUTLINE

1. Follow the Instruction Routine outlined in "To the Laboratory Instructor."

2. Discuss which compounds and ions contribute to the acidity/basicity of a solution and why.

3. Discuss pH. This experiment may be their first exposure to the formal definition and application of pH.

4. Explain, with equations, how ions can produce acidic/basic solutions.

5. Describe or demonstrate the litmus test for testing the acidity/basicity of a solution, Technique 17B.

6. **Part B.3.** Demonstrate the placement of a glowing, not burning, splint in the ammonia-filled bottle.

7. If students are to complete the entire Experimental Procedure, divide the class into three groups to reduce congestion at the reagent table:
   Group I. Do Parts A, B, C, D, E in sequence
   Group II. Do Parts C, D, E, A, B, in sequence
   Group III. Do Parts E, A, B, C, D in sequence

8. A 24-well plate is used for a systematic study of several sets of chemical reactions. The procedure is described so as to minimize the transfer of waste chemicals and the cleansing of the plate.

CAUTIONS & DISPOSAL

- Read aloud the *CAUTION* listed at the beginning of the Experimental Procedure.

- A number of strong and concentrated acids and bases are used in this experiment. Caution students to handle these chemicals with care.

- Have a supply of $NaHCO_3$ handy for cleaning up acid and base spills.

- Provide "Waste Acids" and "Waste Bases" containers.

TEACHING HINTS

1. In this experiment various concentrations of acids are required. As a part of the learning process, have students prepare the more dilute acidic solutions from 6 molar solutions.

2. **Part A.1.** Only Mg and Zn react with 6 M HCl and 6 M CH_3COOH; the Cu also reacts with 6 M HNO_3. The reactions of Mg and Zn with the 6 M CH_3COOH is slow.

3. **Part A.2.** The reaction rate is greatest with 3 M HCl, followed by 0.1 M HCl and 0.01 M HCl. The reaction of the Mg in the HCl solutions is more rapid than in the CH_3COOH solutions of like concentrations.

4. **Part A.3.** I^- oxidizes to I_2 with conc H_2SO_4 and the solution turns amber/purple. HI gas is also evolved. Conc H_2SO_4 is an oxidizing acid; no reaction is observed for conc H_3PO_4.

5. **Part A.4.** Only conc H_2SO_4 has the dehydrating property.

6. **Part B.1.** This is an acid-base neutralization reaction. The litmus is blue in the NaOH solution, but turns pink when the base has been neutralized and the solution becomes acidic.

7. **Part B.2.** $CaO(s) + H_2O(l) \rightarrow Ca(OH)_2(s)$
The resulting solution, being basic, turns red litmus blue.

8. **Part B.3.** $NH_4Cl(s) + Ca(OH)_2(s) \rightarrow CaCl_2(s) + NH_3(g) + H_2O(l)$
Demonstrate or explain the technique for testing for the presence of NH_3, using conc HCl suspended from a stirring rod (footnote 1, Figure 13.8).
    $HCl(g) + NH_3(g) \rightarrow NH_4Cl(s)$
NH_3 is not flammable, it does not support combustion.
NH_3 is *very* soluble in water; it forms a basic solution [$NH_3(aq) + H_2O(l) \rightarrow NH_4^+(aq) + OH^-(aq)$] and turns the phenolphthalein pink.

9. **Part C.1.** The pH of the three water samples will vary, depending upon the source and treatment of the water. Refer students to a universal indicator chart or to the universal indicator photo on the color plate for estimating the pH of the suggested samples.

10. **Part D.1.** The pH of 0.1 M NaCl is 7, the pH's of 0.1 M Na_2CO_3 and 0.1 M Na_3PO_4 are basic, and the pH's of 0.1 M NaH_2PO_4, 0.1 M NH_4Cl, and 0.1 M $KAl(SO_4)_2$ are acidic.

11. **Part E.1.** The most drops of 0.2 M NaOH are the needed to reach the phenolphthalein endpoint for 0.1 M H_2SO_4, next for 0.1 M HCl, and the least for 0.1 M CH_3COOH.

CHEMICALS REQUIRED	**Part A**		**Part C**	
	6 M HCl	2 mL	universal indicator (dropper bottle)	6 mL
	6 M HNO$_3$	2 mL	0.10 M HCl	1 mL
	6 M CH$_3$COOH	2 mL	0.00010 M HCl	1 mL
	Mg, Zn, Cu strips	2 cm	0.10 M CH$_3$COOH	1 mL
	Mg strip	6 cm	0.10 M NH$_3$	1 mL
	3 M HCl	2 mL	0.00010 M NaOH	1 mL
	0.1 M HCl	2 mL	0.10 M NaOH	1 mL
	0.01 M HCl	2 mL	vinegar	1 mL
	3 M CH$_3$COOH	2 mL	lemon juice	1 mL
	0.1 M CH$_3$COOH	2 mL	household ammonia	1 mL
	0.01 M CH$_3$COOH	2 mL	detergent solution	1 mL
	NaI(s)	0.1 g	409®	1 mL
	conc H$_3$PO$_4$ (dropper bottle)	0.2 mL	7-UP®	1 mL
	conc H$_2$SO$_4$ (dropper bottle)	1 mL		
	sugar	1 g	**Part D**	
	conc HCl (dropper bottle)	0.2 mL	0.10 M NaCl	1 mL
			0.10 M Na$_2$CO$_3$	1 mL
	Part B		0.10 M Na$_3$PO$_4$	1 mL
	1 M NaOH	1 mL	0.10 M NaH$_2$PO$_4$	1 mL
	6 M HCl (dropper bottle)	0.5 mL	0.10 M NH$_4$Cl	1 mL
	CaO(s) (freshly opened can)	2 g	0.10 M KAl(SO$_4$)$_2$	1 mL
	NH$_4$Cl(s)	4 g		
	Ca(OH)$_2$(s)	8 g	**Part E**	
	phenolphthalein (dropper bottle)		0.1 M HCl	1 mL
			0.1 M H$_2$SO$_4$	1 mL
			0.1 M CH$_3$COOH	1 mL
			phenolphthalein (dropper bottle)	
			0.2 M NaOH	2 mL

SPECIAL EQUIPMENT				
	24-well plate and Beral pipets	1	rubber tubing	30 cm
	steel wool		gas collecting bottles and glass plates	4
	litmus paper (red/blue)	1 vial of each	800-mL beaker	1
	wood splints	3	universal indicator color chart	
	evaporating dish	1	ring stands and test tube clamps	2
	200-mm test tube	1	rubber tubing	
	balance (±0.01 g)		Bunsen burner	
	gas delivery tube w/rubber stopper for		"Waste Acids" container	
	200-mm test tube	1	"Waste Bases" container	

PRELABORATORY ASSIGNMENT

1. a. Acids turn blue litmus red, have a sour taste, and produce a pricking sensation on skin.

    b. Bases turn red litmus blue, have a bitter taste, and produce a slimy (soapy) feeling on skin.

2. Phenolphthalein is pink in a basic solution and colorless in an acidic solution.

3. The pH number represents the acidity of a solution. A low pH value indicates that the solution has a high molar concentration of H$_3$O$^+$.

    a. Solution 1 has the highest H$_3$O$^+$ molar concentration

    b. Solution 1 is most acidic.

    c. Solution 2 is the most basic.

4. a. Fe^{3+} is more strongly hydrated than Fe^{2+} because the Fe^{3+} has a greater charge density, causing a stronger attraction of adjacent polar water molecules. This stronger attraction weakens the O–H bond in water releasing H$^+$ to the bulk of the solution:
    $$Fe^{3+}\text{---}OH_2(aq) + H_2O(l) \rightleftharpoons FeOH^{2+}(aq) + H_3O^+(aq)$$

    b. Li$^+$ is more strongly hydrated than K$^+$ because of its greater charge density (same charge, but smaller ion). Li$^+$ attracts adjacent water molecules so strongly that the O–H bond is water is weakened permitting H_3O^+ to be released in solution:

$Li^+\text{---}OH_2(aq) + H_2O(l) \rightleftharpoons LiOH(aq) + H_3O^+(aq)$

   c. Fe^{2+}, Cr^{3+} and Cu^{2+} are cations that react with water to produce an acidic solution; thus they are *not* spectator ions.

5. A stirring rod dipped in conc HCl is inserted into the NH_3-filled flask. If the conc HCl "smokes", due to the formation of white NH_4Cl, then NH_3 is present. See footnote 1 in the Experimental Procedure.

6. a.   Limewater is a saturated $Ca(OH)_2$ solution.
    b.   Slaked lime is solid $Ca(OH)_2$.

LABORATORY QUESTIONS

1. a.   The pH will increase because the H_3O^+ concentration decreases and the OH^- concentration increases.
    b.   The pH will decrease because the OH^- concentration decreases and the H_3O^+ concentration increases.

2. a. NH_4^+ and Al^{3+} are cations that produce an acidic solution in Part D.
    b. CO_3^{2-} and PO_4^{3-} are anions that produce a basic solution in Part D.
    c.   Ions that do not affect the pH of the solution are Na^+, K^+, Cl^-, and SO_4^{2-}

*3. $C_{12}H_{22}O_{11}(s) \xrightarrow{\text{conc } H_2SO_4} 12\,C(s) + 11\,H_2O(l)$

*4. $CaO(s) + CO_2(g) \rightarrow CaCO_3(s)$
    $CaO(s) + H_2O(l) \rightarrow Ca(OH)_2(s)$
    and $Ca(OH)_2(s) + CO_2(g) \rightarrow CaCO_3(s) + H_2O(l)$

LABORATORY QUIZ

1. Is an aqueous solution of each of the following acidic, basic, or neutral?
    a.   CaO   d. Na_2O
    b.   NaCl   e. Na_2CO_3
    c. NH_4Cl f. CO_2
       [Answer: acidic: NH_4Cl, CO_2; basic: CaO, Na_2O, Na_2CO_3; neutral: NaCl]

2. Write formulas for
    a.   muriatic acid   d.   lye
    b.   baking soda   e.   quicklime
    c.   milk of magnesia   f.   oil of vitriol
       [Answer: HCl, $NaHCO_3$, $Mg(OH)_2$, NaOH, CaO, H_2SO_4]

3. Which chemical is ranked *first* in production in the United States?   [Answer: H_2SO_4]

4. Define pH.   [Answer: $-\log [H_3O^+]$]

5. Acids turn _____ litmus _____ .   [Answer: blue, red]
    Bases turn phenolphthalein from _____ to _____ .   [Answer: pink, colorless]

6. Which acid provides the most H_3O^+ and which acid provides the least H_3O^+?
    a.   0.1 *M* HCl   c.   0.1 *M* HNO_3
    b.   0.1 *M* H_2SO_4 d.   0.1 *M* CH_3COOH [Answer: H_2SO_4; CH_3COOH]

7. Which cation is more strongly hydrated, Cu^+ of Cu^{2+}? Explain.   [Answer: Cu^{2+}]

8. Why is the pH of an aqueous solution of $CuCl_2$ slightly acidic, but the pH of a $BaCl_2$ solution is neutral?   [Answer: Cu^{2+} reacts with water, Ba^{2+} does not]

Oxidation–Reduction Equations

INTRODUCTION

Students discover that writing and balancing of redox equations are difficult tasks, but with practice they soon begin to enjoy the challenge. However, practice is the key word. This dry lab describes a simple seven-step procedure for balancing redox equations by the ion-electron method in both acidic and basic solutions.

WORK ARRANGEMENT

Individuals.

TIME REQUIREMENT

1 to 2 hours, depending upon your assignment. A suggestion is to assign a select number of the redox equations for balancing over a period of several weeks.

LECTURE OUTLINE

1. Follow the Instruction Routine outlined in "To the Laboratory Instructor."

2. Clearly define and distinguish the terms oxidation, reduction, oxidizing agent, and reducing agent. Use an appropriate example equation.

3. Discuss the seven-step procedure for balancing a redox equation using an appropriate example, *especially* if the procedure has not been discussed in lecture. Balancing a redox equation in both an acidic and a basic solution is suggested.

4. Make the appropriate assignments.

A. Dry Lab Questions

1. a. one H_2O, one H^+
   b. two OH^-, one-half H_2O

2. a. Al, Ag^+
   b. Ag^+, Al

B. Balancing Oxidation-Reduction Equations (Ion-Electron Method)

1. x 3 $(H_2S \rightarrow S + 2H^+ + 2e^-)$
   x 2 $(3e^- + 4H^+ + NO_3^- \rightarrow NO + 2H_2O)$

   $2H^+ + 3\underline{H_2S} + 2\boxed{NO_3^-} \rightarrow 2NO + 3S + 4H_2O$

2. x 2 $(5e^- + 8H^+ + MnO_4^- \rightarrow Mn^{2+} + 4H_2O)$
   x 5 $(2Cl^- \rightarrow Cl_2 + 2e^-)$

   $16H^+ + 2\boxed{MnO_4^-} + 10\underline{Cl^-} \rightarrow 5Cl_2 + 2Mn^{2+} + 8H_2O$

3. x 3 $(2Cl^- \rightarrow Cl_2 + 2e^-)$
   $6e^- + 14H^+ + Cr_2O_7^{2-} \rightarrow 2Cr^{3+} + 7H_2O$

   $14H^+ + 6\underline{Cl^-} + \boxed{Cr_2O_7^{2-}} \rightarrow 2Cr^{3+} + 3Cl_2 + 7H_2O$

4. x 3 $(4Cl^- + HgS \rightarrow HgCl_4^{2-} + S + 2e^-)$
   x 2 $(3e^- + 4H^+ + NO_3^- \rightarrow NO + 2H_2O)$

   $8H^+ + 3\underline{HgS} + 2\boxed{NO_3^-} + 12Cl^- \rightarrow 3HgCl_4^{2-} + 2NO + 3S + 4H_2O$

5. x 5 $(Fe^{2+} \rightarrow Fe^{3+} + e^-)$

$$5\,e^- + 8\,H^+ + MnO_4^- \rightarrow Mn^{2+} + 4\,H_2O$$

$$8\,H^+ + 5\,\underline{Fe^{2+}} + \boxed{MnO_4^-} \rightarrow 5\,Fe^{3+} + Mn^{2+} + 4\,H_2O$$

6. **x 5** $(2\,e^- + 6\,H^+ + BiO_3^- \rightarrow Bi^{3+} + 3\,H_2O)$

    x 2 $(4\,H_2O + Mn^{2+} \rightarrow MnO_4^- + 8\,H^+ + 5\,e^-)$

$$14\,H^+ + 5\,\boxed{BiO_3^-} + 2\,\underline{Mn^{2+}} \rightarrow 2\,MnO_4^- + 5\,Bi^{3+} + 7\,H_2O$$

7. $2\,e^- + 2\,H^+ + 3\,WO_3 \rightarrow W_3O_8 + H_2O$

          $6\,Cl^- + Sn^{2+} \rightarrow SnCl_6^{2-} + 2\,e^-$

$$2\,H^+ + 3\,\boxed{WO_3} + \underline{Sn^{2+}} + 6\,Cl^- \rightarrow W_3O_8 + SnCl_6^{2-} + H_2O$$

8. $2\,e^- + 4\,H^+ + NiO_2 \rightarrow Ni^{2+} + 2\,H_2O$

    x 2 $(Ag \rightarrow Ag^+ + e^-)$

$$4\,H^+ + \boxed{NiO_2} + 2\,\underline{Ag} \rightarrow Ni^{2+} + 2\,Ag^+ + 2\,H_2O$$

9. **x 2** $(e^- + Fe^{3+} \rightarrow Fe^{2+})$

          $H_2S \rightarrow S + 2\,H^+ + 2\,e^-$

$$2\,\boxed{Fe^{3+}} + \underline{H_2S} \rightarrow 2\,Fe^{2+} + S + 2\,H^+$$

10. **x 5** $(2\,e^- + 4\,H^+ + PbO_2 \rightarrow Pb^{2+} + 2\,H_2O)$

     x 2 $(4\,H_2O + Mn^{2+} \rightarrow MnO_4^- + 8\,H^+ + 5\,e^-)$

$$4\,H^+ + 5\,\boxed{PbO_2} + 2\,\underline{Mn^{2+}} \rightarrow 5\,Pb^{2+} + 2\,MnO_4^- + 2\,H_2O$$

11. **x 3** $(5\,e^- + 8\,H^+ + TcO_4^- \rightarrow Tc^{2+} + 4\,H_2O)$

     x 5 $(Ti \rightarrow Ti^{3+} + 3\,e^-)$

$$24\,H^+ + 3\,\boxed{TcO_4^-} + 5\,\underline{Ti} \rightarrow 3\,Tc^{2+} + 5\,Ti^{3+} + 12\,H_2O$$

12. $10\,e^- + 12\,H^+ + 2\,IO_3^- \rightarrow I_2 + 6\,H_2O$

    x 5 $(2\,I^- \rightarrow I_2 + 2\,e^-)$

$$12\,H^+ + 2\,IO_3^- + 10\,I^- \rightarrow 6\,I_2 + 6\,H_2O$$
$$or\ 6\,H^+ + \boxed{IO_3^-} + 5\,\underline{I^-} \rightarrow 3\,I_2 + 3\,H_2O$$

13. **x 5** $(4\,H_2O + C_2H_4 \rightarrow 2\,CO_2 + 12\,H^+ + 12\,e^-)$

   x 12 $(5\,e^- + 8\,H^+ + MnO_4^- \rightarrow Mn^{2+} + 4\,H_2O)$

$$36\,H^+ + 5\,\underline{C_2H_4} + 12\,\boxed{MnO_4^-} \rightarrow 12\,Mn^{2+} + 10\,CO_2 + 28\,H_2O$$

14. **x 4** $(Zn \rightarrow Zn^{2+} + 2\,e^-)$

         $8\,e^- + 8\,H^+ + H_2SO_4 \rightarrow H_2S + 4\,H_2O$

$$8\,H^+ + 4\,\underline{Zn} + \boxed{H_2SO_4} \rightarrow 4\,Zn^{2+} + H_2S + 4\,H_2O$$

15. **x 2** $(8\,OH^- + Cr^{3+} \rightarrow CrO_4^{2-} + 4\,H_2O + 3\,e^-)$

    x 3 $(2\,e^- + 2\,H_2O + MnO_2 \rightarrow Mn^{2+} + 4\,OH^-)$

$$4\,OH^- + 2\,\underline{Cr^{3+}} + 3\,\boxed{MnO_2} \rightarrow 3\,Mn^{2+} + 2\,CrO_4^{2-} + 2\,H_2O$$

16. **x 2** $(3\,e^- + Bi(OH)_3 \rightarrow Bi + 3\,OH^-)$

    x 3 $(2\,OH^- + SnO_2^{2-} \rightarrow SnO_3^{2-} + H_2O + 2\,e^-)$

$$2\,\boxed{Bi(OH)_3} + 3\,\underline{SnO_2^{2-}} \rightarrow 2\,Bi + 3\,SnO_3^{2-} + 3\,H_2O$$

17. **x 2** $(MnO_4{}^{2-} \rightarrow MnO_4{}^- + e^-)$

$$2\,e^- + 2\,H_2O + MnO_4{}^{2-} \rightarrow MnO_2 + 4\,OH^-$$

$$2\,H_2O + 3\,\boxed{MnO_4{}^{2-}} \rightarrow 2\,MnO_4{}^- + MnO_2 + 4\,OH^-$$

18. $\quad 4\,OH^- + Mn^{2+} \rightarrow MnO_2 + 2\,H_2O + 2\,e^-$

$$2\,e^- + Br_2 \rightarrow 2\,Br^-$$

$$4\,OH^- + \underline{Mn^{2+}} + \boxed{Br_2} \rightarrow MnO_2 + 2\,Br^- + 2\,H_2O$$

19. **x 2** $(8\,OH^- + Cr^{3+} \rightarrow CrO_4{}^{2-} + 4\,H_2O + 3\,e^-)$

$\quad$ **x 3** $\underline{(2\,e^- + H_2O_2 \rightarrow 2\,OH^-)}$

$$10\,OH^- + 2\,\underline{Cr^{3+}} + 3\,\boxed{H_2O_2} \rightarrow 2\,CrO_4{}^{2-} + 8\,H_2O$$

20. $\quad 2\,e^- + H_2O + Ag_2O \rightarrow 2\,Ag + 2\,OH^-$

$$3\,OH^- + HPO_3{}^{2-} \rightarrow PO_4{}^{3-} + 2\,H_2O + 2\,e^-$$

$$OH^- + \boxed{Ag_2O} + \underline{HPO_3{}^{2-}} \rightarrow 2\,Ag + PO_4{}^{3-} + H_2O$$

21. **x 3** $(2\,OH^- + NO_2{}^- \rightarrow NO_3{}^- + H_2O + 2\,e^-)$

$\quad$ **x 2** $\underline{(3\,e^- + 2\,H_2O + MnO_4{}^- \rightarrow MnO_2 + 4\,OH^-)}$

$$H_2O + 3\,\underline{NO_2{}^-} + 2\,\boxed{MnO_4{}^-} \rightarrow 3\,NO_3{}^- + 2\,MnO_2 + 2\,OH^-$$

22. $\quad 12\,e^- + 12\,H_2O + P_4 \rightarrow 4\,PH_3 + 12\,OH^-$

$\quad$ **x 3** $\underline{(8\,OH^- + P_4 \rightarrow 4\,H_2PO_2{}^- + 4\,e^-)}$

$$12\,H_2O + 12\,OH^- + 4\,P_4 \rightarrow 4\,PH_3 + 12\,H_2PO_2{}^-$$
$$\textit{or } 3\,H_2O + 3\,OH^- + \boxed{P_4} \rightarrow PH_3 + 3\,H_2PO_2{}^-$$

23. $\quad 2\,OH^- + CN^- \rightarrow OCN^- + H_2O + 2\,e^-$

$\quad$ **x 2** $\underline{(e^- + [Fe(CN)_6]^{3-} \rightarrow [Fe(CN)_6]^{4-})}$

$$2\,OH^- + \underline{CN^-} + 2\,\boxed{[Fe(CN)_6]^{3-}} \rightarrow OCN^- + 2\,[Fe(CN)_6]^{4-} + H_2O$$

24. **x 3** $(8\,OH^- + N_2H_4 \rightarrow 2\,NO + 6\,H_2O + 8\,e^-)$

$\quad$ **x 4** $\underline{(6\,e^- + 3\,H_2O + ClO_3{}^- \rightarrow Cl^- + 6\,OH^-)}$

$$3\,\underline{N_2H_4} + 4\,\boxed{ClO_3{}^-} \rightarrow 6\,NO + 4\,Cl^- + 6\,H_2O$$

25. $\quad 2\,e^- + S_2O_8{}^{2-} \rightarrow 2\,SO_4{}^{2-}$

$$2\,OH^- + Ni(OH)_2 \rightarrow NiO_2 + 2\,H_2O + 2\,e^-$$

$$2\,OH^- + \boxed{S_2O_8{}^{2-}} + \underline{Ni(OH)_2} \rightarrow 2\,SO_4{}^{2-} + NiO_2 + 2\,H_2O$$

Oxidation-Reduction Reactions

INTRODUCTION	Most chemical reactions are classified as either acid-base or redox. While Experiment 13 covers the principles and observations associated with acid-base reactions, this experiment covers that of redox reactions. Dry Lab 2A reviews the oxidation numbers of elements in compounds and Dry Lab 4 covers a systematic procedure for the balancing of redox equations. An understanding of both concepts are of value in obtaining a full appreciation of this experiment.

WORK ARRANGEMENT	Partners. Students learn from the shared experience.

TIME REQUIREMENT	2.5 hours

LECTURE OUTLINE

1. Follow the Instruction Routine outlined in "To the Laboratory Instructor."

2. Review the definitions, with examples, of oxidizing and reducing agents.

3. If Dry Lab 4, was not assigned, then some time should be spent in explaining how to balance redox equations. An example from a reaction in Part A.3 is appropriate.

4. Students create an abbreviated activity series in Parts B and C through a series of observed single displacement reactions. A short discussion of an activity series may be helpful to students.

CAUTIONS & DISPOSAL

• Dispose of the test solutions in the "Waste Salts" container and the unreacted metals in the "Waste Solids" container.

• **Part A.1.** Advise students *not* to look at the burning Mg ribbon.

• **Part A.2** recommends the use of 6 M HCl (also Parts A.3 and B.1), 6 M HNO_3, and conc HNO_3; **Part A.3** recommends the use of 6 M H_2SO_4. Warn students about the mishandling of strong acids.

• Students should wash their hands before leaving the laboratory.

TEACHING HINTS

1. **Part A.1.** Do *not* let students play with the magnesium ribbon. The equation for the ignition of magnesium ribbon in air is: $2\,Mg(s) + O_2(g) \rightarrow 2\,MgO(s)$

2. **Part A.2.** Copper metal reacts more vigorously with conc HNO_3 than with the 6 M HNO_3; Cu does *not* react with the 6 M HCl. Students may have difficulty in distinguishing between the two reactions of Cu with HNO_3: with conc HNO_3, the red-brown $NO_2(g)$ is a direct product of the reaction but with 6 M HNO_3, $NO_2(g)$ is a secondary product—the $NO(g)$ product quickly air oxidizes to NO_2. Close observations can make the distinction however.

   For 6 M HNO_3: $3\,Cu(s) + 8\,HNO_3(aq) \rightarrow 3\,Cu(NO_3)_2(aq) + 2\,NO(g) + 4\,H_2O(l)$
   For conc HNO_3: $Cu(s) + 4\,HNO_3(aq) \rightarrow Cu(NO_3)_2(aq) + 2\,NO_2(g) + 2\,H_2O(l)$

3. A 24-well plate is the apparatus for testing in Parts A.3 through C.

4. **Part A.3.** A number of redox reactions are observed and the balanced equations are written. Students may need some assistance or guidance in making their observations and writing their balanced equations. The observations and the balanced equations for the reactions are:

Well A1   Formation of a deep blue $I_2 \cdot$starch complex.

$$6\,I^-(aq) + NO_3^-(aq) + 8\,H^+(aq) \xrightarrow{\text{starch}} 3\,I_2\cdot\text{starch}(aq) + 2\,NO(g) + 4\,H_2O(l)$$

Well A2   Discoloration of the purple permanganate ion.

$$MnO_4^-(aq) + 5\,Fe^{2+}(aq) + 8\,H^+(aq) \rightarrow 5\,Fe^{3+}(aq) + Mn^{2+}(aq) + 4\,H_2O(l)$$

Well A3   Discoloration of permanganate ion, the formation of the *very* light pink manganese(II) ion, and the evolution of a gas (may be difficult to observe). The reaction is slow to occur.

$$2\,MnO_4^-(aq) + 5\,C_2O_4^{2-}(aq) + 16\,H^+(aq) \rightarrow 2\,Mn^{2+}(aq) + 10\,CO_2(g) + 8\,H_2O(l)$$

Well A4   Discoloration of the purple permanganate solution.

$$2\,MnO_4^-(aq) + 5\,HSO_3^-(aq) + H^+(aq) \rightarrow 2\,Mn^{2+}(aq) + 5\,SO_4^{2-}(aq) + 3\,H_2O(l)$$

Well B1   Formation of deep blue $I_2\cdot$starch complex.

$$ClO^-(aq) + 2\,I^-(aq) + 2\,H^+(aq) \xrightarrow{\text{starch}} I_2\cdot\text{starch}(aq) + Cl^-(aq) + H_2O(l)$$

Well B2   Discoloration of the orange dichromate solution.

$$Cr_2O_7^{2-}(aq) + 6\,Fe^{2+}(aq) + 14\,H^+(aq) \rightarrow 6\,Fe^{3+}(aq) + 2\,Cr^{3+}(aq) + 7\,H_2O(l)$$

Well B3   Discoloration of the orange dichromate ion, the formation of the light blue chromium (III) ion and the evolution of a gas (may be difficult to observe).

$$Cr_2O_7^{2-}(aq) + 3\,C_2O_4^{2-}(aq) + 14\,H^+(aq) \rightarrow 2\,Cr^{3+}(aq) + 6\,CO_2(g) + 7\,H_2O(l)$$

Well B4   Discoloration of the orange dichromate solution and the formation of the light blue chromium (III) solution.

$$Cr_2O_7^{2-}(aq) + 3\,NO_2^-(aq) + 8\,H^+(aq) \rightarrow 2\,Cr^{3+}(aq) + 3\,NO_3^-(aq) + 4\,H_2O(l)$$

5. **Part B.1.** Each of the metals should be thoroughly cleaned with steel wool (esp. Al and Mg) before being added to the 6 *M* HCl. Ni(slow), Zn, Fe, Al, and Mg react with the H_3O^+; Cu does not.

6. **Part C.** Again, each of the metals should be polished with steel wool. Evidence of reaction may be slight; the appearance of a dull finish on a freshly polished surface is evidence of displacement. The observations that should be observed are summarized:

	Ni	Cu	Zn	Fe	Al	Mg
HCl	r	nr	r	r	r	r
NiSO$_4$	—	nr	r	r		
Cu(NO$_3$)$_2$	r	—	r	r		
Zn(NO$_3$)$_2$	nr	nr	—	nr		
Fe(NH$_4$)$_2$(SO$_4$)$_2$	nr	nr	r	—		

The abbreviated activity series should read Mg > Al > Zn > Fe > Ni > Cu

7. **Part C.2.** You may choose to add other metals to the investigation of chemical reactivity. Metals such as silver and lead are suggested.

CHEMICALS REQUIRED				
Mg ribbon	2 cm	0.1 *M* Fe(NH$_4$)$_2$(SO$_4$)$_2$	3.5 mL
Cu wire/strip	3 cm	1 *M* K$_2$C$_2$O$_4$ (dropper bottle)	0.5 mL
6 *M* HCl	10 mL	0.1 *M* NaHSO$_3$ (dropper bottle)	0.5 mL
6 *M* HNO$_3$	1 mL	5% NaClO (chlorine bleach)	2 mL
conc HNO$_3$	1 mL	0.01 *M* K$_2$Cr$_2$O$_7$	3 mL
0.1 *M* HNO$_3$	1 mL	KNO$_2$(s)	0.1 g
0.1 *M* KI	1 mL	Ni, Cu, Zn, Fe, Al, Mg strips	1 cm
starch solution (dropper bottle)	0.5 mL	0.1 *M* NiSO$_4$	3 mL
0.01 *M* KMnO$_4$	3 mL	0.1 *M* Cu(NO$_3$)$_2$	3 mL
6 *M* H$_2$SO$_4$ (dropper bottle)	1 mL	0.1 *M* Zn(NO$_3$)$_2$	3 mL

SPECIAL EQUIPMENT	24-well plate and Beral pipets steel wool litmus (red/blue) 　　　　　1 vial of each	Bunsen burner "Waste Salts" container "Waste Solids" container

PRELABORATORY ASSIGNMENT

1. a. An oxidizing agent is a chemical that causes oxidation of another substance and, in turn, accepts electrons.
 b. A reducing agent is a chemical that provides electrons for the reduction of another substance, thereby losing electrons.
 c. A half-reaction represents either the oxidation or reduction part of a redox reaction.

2. No. The equation is balanced for mass but *not* for charge. The balanced equation is:
$$6\,H^+(aq) + 2\,MnO_4^-(aq) + 5\,H_2C_2O_4(aq) \rightarrow 2\,Mn^{2+}(aq) + 10\,CO_2(g) + 8\,H_2O(l)$$

3. (most reactive) X > R > T > M > Q > Z (least reactive)

*4. Zn is more reactive than Fe; therefore in an oxidation reaction with air, the zinc is preferentially oxidized to the more structurally sound Fe.

LABORATORY QUESTIONS

1. *No*. Cu metal is not oxidized by HCl; HNO_3, an oxidizing acid, oxidizes copper metal, producing NO (or NO_2) gas and copper(II) ion.

2. Sodium metal would be at the (*very*) active end of the activity series; gold would be placed at the nonreactive end.

3. Mg metal and Cu^{2+} ion.

*4. When aluminum and magnesium react with oxygen, a tough oxide coating forms on the metal surface. This coating is impervious to further air oxidation of the lower layers of the metal. This is especially so for aluminum metal.

*5. Mg is more reactive than Fe; therefore in an oxidation reaction with air and water in the soil, the magnesium is preferentially oxidized instead of the more structurally sound Fe.

LABORATORY QUIZ

1. Zinc metal is more reactive than nickel metal. Does a reaction occur when a strip of zinc metal is placed into a $NiCl_2$ solution? If "yes", then write the balanced net ionic equation for the reaction.　　[Answer: Yes; $Zn(s) + Ni^{2+}(aq) \rightarrow Zn^{2+}(aq) + Ni(s)$]

2. When chlorine gas oxidizes manganese(II) oxide to permanganate ions in a basic solution, chloride ions form. Write a balanced equation.
 [Answer: $3\,Cl_2(g) + 8\,OH^-(aq) + 2\,MnO_2(s) \rightarrow 6\,Cl^-(aq) + 2\,MnO_4^-(aq) + 4\,H_2O(l)$]

3. If mercury metal is more reactive than platinum metal, which metal is the better reducing agent? Explain.　　　　　　　　　　　　　　　　　　[Answer: Hg]

4. Mercury metal does not react with hydrochloric acid but it reacts with nitric acid.
 a. What is the most probable oxidation state of mercury in solution? [Answer: Hg^{2+}]
 b. Is mercury metal or nitric acid the oxidizing agent?　　　　[Answer: HNO_3]
 c. Since HCl and HNO_3 are acids, why does mercury metal react with nitric acid and not hydrochloric acid?　　　　　[Answer: HNO_3 is an oxidizing acid, HCl is not]

5. Given the following activity series: Ba > Mn > Cr > Cd > Sn > H > Bi > Au
 Predict which of the following reactions occur:
 a. $3\,Mn(s) + 2\,Cr^{3+}(aq) \rightarrow 3\,Mn^{2+}(aq) + 2\,Cr(s)$　　　　　　　[Yes]
 b. $2\,Bi(s) + 6\,H^+(aq) \rightarrow 3\,H_2(g) + 2\,Bi^{3+}(aq)$　　　　　　　　[No]
 c. $3\,Cd^{2+}(aq) + 2\,Au(s) \rightarrow 3\,Cd(s) + 2\,Au^{3+}(aq)$　　　　　　[No]
 d. $Sn(s) + 2\,H^+(aq) \rightarrow H_2(g) + Sn^{2+}(aq)$　　　　　　　　　[Yes]
 e. $2\,Cr^{3+}(aq) + 3\,Ba(s) \rightarrow 3\,Ba^{2+}(aq) + 2\,Cr(s)$　　　　　　[Yes]
 f. $2\,Au^{3+}(aq) + 3\,H_2(g) \rightarrow 6\,H^+(aq) + 2\,Au(s)$　　　　　　　[Yes]
 g. $3\,Mn^{2+}(aq) + 2\,Bi(s) \rightarrow 3\,Mn(s) + 2\,Bi^{3+}(aq)$　　　　　　[No]

Bleach Analysis

INTRODUCTION

The oxidizing strength of household bleach is determined. The hypochlorite ion, ClO^-, is the oxidizing agent. An iodometric titration is used to determine the moles of ClO^- present in the sample, calculated as "available chlorine" in a bleaching agent. Iodine, I_2, generated from the reaction between the ClO^- in the bleach and I^-, is titrated with a standardized $Na_2S_2O_3$ solution.

This experiment follows the qualitative redox experiment (Experiment 14). A review of volumetric analyses introduced in Experiment 9 is appropriate in introducing this experiment. You and students should carefully review Technique 16.

The procedure for an iodometric analysis requires some understanding of the chemical system before entering the laboratory. A good lab preparation for you as a lab instructor is to work through the Prelaboratory Assignment. In terms of an understanding, this experiment is perhaps the most difficult at this point in the manual.

Do not plan to give a quiz during this lab period—it will take *at least* 3 hours for most students to complete the experiment. Allowing 2 laboratory periods for the completion of all parts of the experiment is also suggested.

WORK ARRANGEMENT

Individuals.

TIME REQUIREMENT

3 hours; 1.5 hours for Parts A and B and 1.5 hours for Parts C (or D) and E. Additional time is required for the completion of the calculations. To save time the standardized $Na_2S_2O_3$ solutions can be prepared in advance by stockroom personnel or have students analyze only one bleach sample instead of two.

LECTURE OUTLINE

1. Follow the Instruction Routine outlined in "To the Laboratory Instructor."

2. Students do not always clearly understand iodometric titrations, especially the reason for the formation of the I_2•starch complex (Equation 15.4) and its dissociation for the analysis (Equation 15.5). Review in detail Equations 15.3 to 15.6. From Equation 15.6, explain how the "available chlorine" is calculated for the bleach sample.

3. Discuss the standardization procedure for the $Na_2S_2O_3$ solution. The analysis should be done in triplicate.

4. Prelaboratory calculations in Part A.1 for the preparation of 100 mL of a 0.01 M KIO_3 solution and in Part B.1 for the preparation of 250 mL of a 0.1 M $Na_2S_2O_3$ solution are required.

5. In Parts B and E, the addition of starch should be delayed until the stoichiometric point is near, or when only a faint yellow of I_2 remains in solution. You may choose to demonstrate this to the class (now, or with a student's titration in the laboratory). The I_2•starch complex is slow to dissociate and, therefore, the stoichiometric point can be easily surpassed (the most common experimental error in the experiment) if the starch is added too early.

CAUTIONS & DISPOSAL

- The unused KIO_3 solution should be discarded in the "Waste Oxidants" container.

- The unused $Na_2S_2O_3$ solution should be discarded in the "Waste Thiosulfate" container.

- The unused H_2SO_4 solution should be discarded in the "Waste Acids" container.

TEACHING HINTS	1. **Part A.1.** 0.214 g of KIO_3 (molar mass = 214.0 g/mol) is required to prepare 100 mL of 0.01 M KIO_3.
	2. **Part B.1.** 6.20 g of $Na_2S_2O_3 \cdot 5H_2O$ (molar mass = 248.2 g/mol) is required to prepare 250 mL of 0.1 M $Na_2S_2O_3$.
	3. **Part B.3.** Supervise the pipetting and titration techniques, and the addition of the starch near the stoichiometric point.
	4. **Part B.4.** Good precision should be required for the standardization of the $Na_2S_2O_3$ solution.
	5. **Part D.** In analyzing a powdered bleach sample, the starch can be added immediately after KI addition. Only a small amount of $Na_2S_2O_3$ is required to reach the endpoint.
	6. **Parts B and E.** Near the stoichiometric point, $Na_2S_2O_3$ addition should be slowed, even stopped on occasion, to allow time for the dissociation of the $I_2 \cdot$starch complex.
	7. Some assistance in the calculations may be required.

CHEMICALS REQUIRED	*$KIO_3(s)$	0.3 g	0.5 M H_2SO_4	70 mL
	**$Na_2S_2O_3 \cdot 5H_2O(s)$	6.5 g	starch solution	15 mL
	$KI(s)$	14 g		

*Dry in a 110°C drying oven 24 hours before the laboratory period and store in a desiccator.
**To expedite the analyses, prepare 250 mL of 0.1 M $Na_2S_2O_3$ (per student) several days in advance.

SUGGESTED UNKNOWNS	Liquid bleach, commercial: a 5% $Ca(ClO)_2$ solution, or a 5.3% NaClO solution	15 mL	Mildew remover is a 4.8% $Ca(ClO)_2$ solution	25 mL
	Laundry bleach, solid	5 g	Commercial HTH, used for home swimming pools, is 65% $Ca(ClO)_2$	
	Powdered bleach: commercial or 0.43% $Ca(ClO)_2$	5 g		

SPECIAL EQUIPMENT	weighing paper	5	mortar and pestle	1
	balance (±0.001 g)		weighing paper	5
	drying oven		ring stand and buret clamp	
	desiccator or desicooler		"Waste Oxidants" container	
	100-mL volumetric flask	2	"Waste Thiosulfate" container	
	50-mL buret and buret brush	1	"Waste Acids" container	
	25-mL pipet and pipet bulb	1	"Waste Solids" container	
	10-mL pipet and pipet bulb	1		

PRELABORATORY ASSIGNMENT	1. a. "Available chlorine" is the mass of oxidizing agent in a measured mass of a sample of bleach that is equivalent to an equal mass of Cl_2.
	b. According to Equations 15.3 and 15.6, one mole of ClO^- and one mole of Cl_2 each react with two moles of I^-; therefore, each mole of ClO^- makes available one mole of Cl_2.
	2. a. blue to colorless
	b. the blue $I_2 \cdot$starch complex dissociates with the addition of the standardized $Na_2S_2O_3$ reagent, Equation 15.5.
	3. The ClO^- ion oxidizes I^- to I_2. The amount of I_2 that is generated by the reaction is then titrated with a standard $S_2O_3^{2-}$ solution.
	4. $Na_2S_2O_3$ is a reducing agent; it reduces I_2 in the $I_2 \cdot$starch complex to I^-.
	$2\ S_2O_3^{2-}(aq) + 2\ e^- \rightarrow S_4O_6^{2-}(aq)$

5. a. $0.0227 \text{ L} \times \dfrac{0.107 \text{ mol}}{\text{L}} \times \dfrac{1 \text{ mol I}_2}{2 \text{ mol S}_2\text{O}_3{}^{2-}} \times \dfrac{1 \text{ mol ClO}^-}{1 \text{ mol I}_2}$

   $= 1.21 \times 10^{-3}$ mol ClO$^-$ in 25 mL titrated sample

   b. 1.21×10^{-3} mol ClO$^- \times \dfrac{1 \text{ mol Cl}_2}{1 \text{ mol ClO}^-} \times \dfrac{71.0 \text{ g Cl}_2}{\text{mol Cl}_2} = 0.0862$ g Cl$_2$

   c. 2.5 mL bleach $\times \dfrac{1.08 \text{ g bleach}}{\text{mL bleach}} = 2.7$ g bleach

   d. 3.19% available Cl$_2$

*6. According to Equation 15.3, an addition of acid shifts the equilibrium left; this increases the concentration of Cl$_2$, a gas with a low solubility in water, which is very toxic.

LABORATORY QUESTIONS

1. The I$_2$•starch complex is slow to dissociate. If starch were added earlier and if ample time were not allowed for the I$_2$•starch to dissociate during the titration, there is a good chance that too much of the Na$_2$S$_2$O$_3$ may be added and the stoichiometric point will be surpassed.

2. *High.* If an air bubble is trapped in the buret tip but then released, it is recorded as volume of S$_2$O$_3{}^{2-}$ titrant added. This implies that more I$_2$ was formed by the ClO$^-$ producing a *larger than actual* percent available Cl$_2$ in the sample.

3. *High.* If the molar concentration of the Na$_2$S$_2$O$_3$ solution is reported too high, then an *apparently* large amount of ClO$^-$ will react resulting in a reported *high* percent available Cl$_2$.

4. a. $0.0217 \text{ L} \times \dfrac{0.100 \text{ mol}}{\text{L}} \times \dfrac{1 \text{ mol I}_2}{2 \text{ mol S}_2\text{O}_3{}^{2-}} \times \dfrac{1 \text{ mol ClO}^-}{1 \text{ mol I}_2} = 1.09 \times 10^{-3}$ mol ClO$^-$.

   1.09×10^{-3} mol ClO$^-$ also equals 1.09×10^{-3} mol Cl$_2$; 7.70×10^{-2} g Cl$_2$

   b. Assuming an undiluted sample for the analysis:

   $\dfrac{7.70 \times 10^{-2} \text{ g Cl}_2}{0.547 \text{ g sample}} \times 100 = 14.1\%$ available Cl$_2$

LABORATORY QUIZ

1. a. What is the oxidizing agent in the equation:
   ClO$^-$(aq) + I$^-$(aq) + H$_2$O(l) → I$_2$(aq) + Cl$^-$(aq) + 2 OH$^-$(aq)   [Answer: ClO$^-$]
   b. What is the oxidizing agent in the equation:
   I$_2$(aq) + 2 S$_2$O$_3{}^{2-}$(aq) → 2 I$^-$(aq) + S$_4$O$_6{}^{2-}$(aq)   [Answer: I$_2$]

2. Chlorine gas oxidizes iodide to iodine in an aqueous solution.
   a. Write a balanced equation.   [Answer: Cl$_2$(g) + 2 I$^-$(aq) → 2 Cl$^-$(aq) + I$_2$(s)]
   b. Calculate the mass of chlorine required to reach the stoichiometric point in titrating 2.67 mg of I$^-$.   [Answer: 0.746 mg Cl$_2$]

3. In the first step of the bleach analysis, the volume of the bleach solution was measured with a pipet; however, the amount of KI added to the bleach was not critical. From the equations in Question 1 above, account for this procedural step.

Stoichiometric Analysis of a Redox Reaction

INTRODUCTION

Depending upon the perceived level of our students and/or the cycle of the semesters we often alternate this experiment with Experiment 15, both being quantitative redox experiments. The equations for the reactions in this experiment, not being precisely defined in the Introduction, are somewhat bewildering to the students and, consequently, often requires some explanation.

The experiment can be best described as "open ended" in that the stoichiometry of the reaction may vary (and the experiment is so written as to imply that) from student to student.

WORK ARRANGEMENT

Individuals.

TIME REQUIREMENT

2.5 hours

LECTURE OUTLINE

1. Follow the Instruction Routine outlined in "To the Laboratory Instructor."

2. Explain in detail how the chemistry described by Equations 16.5 through 16.7 are used to complete Equation 16.4. The chemistry of this experiment is not often clearly understood, even by good chemistry students. Even so, the analysis and the results of the experiment can be very gratifying.

3. The chemistry for the analysis of the reaction system and the standardization of the $KMnO_4$ solution are described in the Introduction and in the Teaching Hints below.

4. Calculations of the mass of $KMnO_4$ in Part A.1, the mass of $Na_2C_2O_4$ in Part A.4, and the volume of 0.2 M $FeNH_4(SO_4)_2$ in Part B.2 are required before the procedure can continue.

5. Students , for the most part, design their own Report Sheet.

CAUTIONS & DISPOSAL

- 6 M H_3PO_4 in Part C.2 should be handled with care.

- Advise students to dispose of the test solutions in the sink, followed by a generous flow of water. The unused $KMnO_4$ solution should be discarded in the "Waste Oxidants" container.

TEACHING HINTS

1. **Part A.1.** A mass of $\approx$0.32 g of $KMnO_4$ is required for the preparation of the 100 mL of 0.020 M $KMnO_4$ solution.

2. **Part A.2.** Boiled, deionized water is used for the preparation of the solution. Wrap the flask in aluminum foil.

2. **Part A.3.** The *top* of the meniscus should be read when $KMnO_4$ is the titrant.

3. **Part A.4.** The balanced equation for the standardization of the $KMnO_4$ solution (Equation 16.11) is:
$$2\ MnO_4^-(aq) + 5\ C_2O_4^{2-}(aq) + 16\ H^+(aq) \rightarrow 2\ Mn^{2+}(aq) + 10\ CO_2(g) + 8\ H_2O(l)$$
A mass of $\approx$0.089 g of $Na_2C_2O_4$ is to be measured for the standardization of the $KMnO_4$ solution.

4. **Part A.5.** The analyte is heated to 80°C to not only drive of the CO_2, but to also increase the reaction rate.

5. **Part B.1.** Use a clean 10-mL pipet and the proper pipetting procedure (Technique 16B).

6. **Part B.2.** About 15 mL of 0.2 M $FeNH_4(SO_4)_2$ is to be measured for the analysis; the NH_3OH^+ reduces the Fe^{3+} to Fe^{2+}. The amount of Fe^{2+} generated in this reaction is determined in Part C.

CHEMICALS REQUIRED	*$KMnO_4$	0.4 g	0.05 M (to 3 sig fig) $NH_3OH^+Cl^-$	40 mL
	$Na_2C_2O_4$ (dried at 110°C)	0.3 g	0.2 M $FeNH_4(SO_4)_2$	60 mL
	0.9 M H_2SO_4	150 mL	6 M H_3PO_4	40 mL

*Light catalyzes the decomposition of $KMnO_4$; store it in a *clean*, brown bottle in a dark place. Standardize the solution (ideally) the day of the experiment.

SPECIAL EQUIPMENT	weighing paper		10-mL pipet and pipet bulb	1
	balance (±0.01 g)		balance (±0.001 g)	
	100-mL volumetric flask	1	125-mL Erlenmeyer flasks	3
	aluminum foil (to cover $KMnO_4$ flask)		ring stand and buret clamp	1
	50-mL buret and buret brush	1	Bunsen burner	
			"Waste Oxidants" container	

PRELABORATORY ASSIGNMENT

2. a. $0.100 \text{ L} \times \dfrac{0.020 \text{ mol } KMnO_4}{L} \times \dfrac{158 \text{ g } KMnO_4}{mol} = 0.316 \text{ g } KMnO_4$

   b. $0.015 \text{ L} \times \dfrac{0.020 \text{ mol } KMnO_4}{L} \times \dfrac{5 \text{ mol } C_2O_4^{2-}}{2 \text{ mol } MnO_4^-} \times \dfrac{118 \text{ g } Na_2C_2O_4}{\text{mol } C_2O_4^{2-}} = 0.089 \text{ g } Na_2C_2O_4$

   c. $0.010 \text{ L} \times \dfrac{0.05 \text{ mol } NH_3OH^+}{L} \times \dfrac{6 \text{ mol } Fe^{3+}}{1 \text{ mol } NH_3OH^+} \times \dfrac{L \text{ solution}}{0.2 \text{ mol } FeNH_4(SO_4)_2}$
   $= 0.015 \text{ L} = 15 \text{ mL of } 0.2 \, M \, FeNH_4(SO_4)_2$

3. Dissolve 0.316 g $KMnO_4$ (see Prelaboratory Question 2.a) in a 100 mL volumetric flask and dilute to the mark with previously boiled, deionized water.

4. $0.01772 \text{ L} \times \dfrac{0.0500 \text{ mol } KMnO_4}{L} \times \dfrac{5 \text{ mol } Fe^{2+}}{1 \text{ mol } MnO_4^-} \times \dfrac{259.9 \text{ g } FeSO_4 \cdot 6H_2O}{mol}$
   $= 1.15 \text{ g } FeSO_4 \cdot 6H_2O$

5. $2 NH_3OH^+ + 2 Fe^{3+} \rightarrow 2 Fe^{2+} + N_2 + 2 H_2O + 4 H^+$
   $2 NH_3OH^+ + 4 Fe^{3+} \rightarrow 4 Fe^{2+} + N_2O + H_2O + 6 H^+$
   $NH_3OH^+ + 3 Fe^{3+} \rightarrow 3 Fe^{2+} + NO + 4 H^+$
   $NH_3OH^+ + 5 Fe^{3+} + H_2O \rightarrow 5 Fe^{2+} + NO_2 + 6 H^+$
   $NH_3OH^+ + 6 Fe^{3+} + 2 H_2O \rightarrow 6 Fe^{2+} + NO_3^- + 8 H^+$

LABORATORY QUESTIONS

1. The mass measurement of the $KMnO_4$ is not critical because it is the $KMnO_4$ solution that is to be standardized.

2. As the analysis is occurring in Part A.5, the oxalate ion is being oxidized to CO_2 which is being evolved with the heat. Mn^{2+} is being generated in the reaction which, with its accumulation, catalyzes the reaction.

3. Fe^{2+} would have been substituted for Fe^{3+} in the mistaken use of the chemical.
   a. None of the NH_3OH^+ would be oxidized (and none of the Fe^{2+} would be reduced) in the reaction mixture.
   b. *All* of the Fe^{2+} that was added in Part B.2 as $Fe(NH_4)_2(SO_4)_2$ would now be oxidized with the standardized MnO_4^- in Part C.3.

4. a. $\dfrac{0.0730 \text{ mol } Fe^{3+}}{0.0146 \text{ mol } NH_3OH^+}$ is a 5:1 mole ratio, indicating a 5 electron exchange between

NH_3OH^+ and Fe^{2+}. Therefore, according to the solution to Prelaboratory Question 5, the oxidation product of NH_3OH^+ must be NO_2.

b. $NH_3OH^+ + H_2O \rightarrow NO_2 + 6\,H^+ + 5\,e^-$

LABORATORY QUIZ 1. What is the color of the iron(II) ion? the iron(III) ion?

2. In dissolving an iron sample with nitric acid, all iron exists as the iron(III) ion in solution. Describe what would happen if an attempt is made to titrate this sample with standardized $KMnO_4$.   [Answer: no reaction would occur]

3. A 1.40-g mixture is 90% $Fe(NH_4)_2(SO_4)_2 \cdot 6H_2O$ and 10% Na_2SO_4 by mass.
    a. What is the mass of iron in the mixture?   [Answer: 0.179 g]
    b. Calculate the volume of 0.020 M $KMnO_4$ required to oxidize the iron(II) ion to the iron(III) ion?   [Answer: 32.1 mL]

4. A 0.115-g sample of $K_2C_2O_4$ requires 15.74 mL of a $KMnO_4$ solution to reach the stoichiometric point. What is the molar concentration of the $KMnO_4$ solution?
                                       [Answer: 0.0195 M $KMnO_4$]

Molar Mass of a Volatile Liquid

INTRODUCTION

The ideal gas law equation and the van der Waal equation are used to determine the molar mass of a volatile organic compound in this experiment. Solving the ideal gas law equation using the data that are collected in the laboratory should be relatively easy for students.

Using the Van der Waal equation requires some advanced mathematical skills; your discretion to incorporate this aspect of the principle of gases into this experiment is required.

The determination of *n* from the van der Waal equation requires the solving of a cubic equation (see Prelab Question 5 for the solution). Since students cannot do this easily, solve the van der Waal equation first by substitution of the "ideal" *n* and then vary *n* until a satisfactory equality results. Students may need some instruction for completing the calculations.

WORK ARRANGEMENT

Partners. This permits discussion and consultation between students.

TIME REQUIREMENT

2.5 hours (0.5 hours for the van der Waal calculation)

LECTURE OUTLINE

1. Follow the Instruction Routine outlined in "To the Laboratory Instructor."

2. Review the ideal gas law and units of molar mass. Identify the data that are to be recorded.

3. Discuss the van der Waal equation and how it is to be used in the experiment. See Prelab Question 5.

4. A collection of class data is suggested for this experiment. For a given organic compound the standard deviation of the collected class data is to be calculated. Students may need some guidance for the collection and analysis of the class data.

CAUTIONS & DISPOSAL

- A slow-boiling water bath is at the same temperature as a fast-boiling water bath—encourage students to *gently* boil the water in Part B.

- The organic liquid is flammable—keep the Bunsen flame away from the flammable organic unknowns.

- Provide a "Waste Organics" container for the disposal of excess sample.

TEACHING HINTS

1. **Parts A.1, 2.** Cover the flask with the aluminum foil and secure it with a rubber band; the aluminum foil and rubber band are to be included in the mass measurements for Parts A and B.

2. **Part B.1.** Neither the Erlenmeyer flask, nor the clamp should touch the beaker wall. The water level should be high on the Erlenmeyer flask. To minimize condensation of the organic vapor on the neck of the Erlenmeyer flask, wrap the top of the beaker and the flask with aluminum foil.

3. **Part B.2.** The organic vapor that escapes through the pinholes of the aluminum foil is difficult to see. Maintaining the boiling water bath for a longer period of time is o.k.; be sure that the water level is maintained in the beaker. *Not* vaporizing the organic liquid is the greatest source of experimental error.

4. **Part B.3.** The outside of the flask must be dry before any mass determination.

5. **Part C.1.** Measure the volume of the flask as accurately as possible with a graduated cylinder.

6. Results are usually within ± 5% using the ideal gas law equation. This improves slightly with the use of the van der Waal equation.

SUGGESTED UNKNOWNS

The organic unknowns should have a boiling point less than 90°C.

methanol	65°C	32.0 g/mol
ethanol (this *must be* absolute ethanol)	78°C	46.1 g/mol
acetone	56 °C	58.1 g/mol
iso-propanol	82°C	60.1 g/mol
hexane	69°C	86.2 g/mol
cyclohexane	81°C	84.2 g/mol
1-pentane	36°C	72.2 g/mol

SPECIAL EQUIPMENT

aluminum foil and rubber band		ring stand and iron support rings	1
balance (±0.001 g)		utility clamp	
600-mL beaker	1	Bunsen burner	
boiling chips		110°C thermometer	1
drying oven (optional)		laboratory barometer	
		"Waste Organics" container	

PRELABORATORY ASSIGNMENT

1. a. 1.00×10^{-2} mol
   b. 77.5 g/mol

2. a. 75.5 g/mol
   b. 2.58 %

3. The mass is determined by subtracting the mass of the flask, aluminum foil, and rubber band from that of the flask, aluminum foil, rubber band, and vaporized unknown.

4. a. 43.2 g/mol
   b. $\pm 0.36 \approx \pm 0.4$

5. (optional) $n^3 \left(\dfrac{ab}{V^2}\right) - n^2 \left(\dfrac{a}{V}\right) + n(Pb + RT) = PV$

LABORATORY QUESTIONS

1. *High.* The mass of the water on the flask is also measured and will be considered as added mass from the unknown. A greater mass causes a *higher* calculated molar mass.

2. a. If the actual pressure is less than 1 atm but assumed to be 1 atm, this infers a greater number of moles and a lower molar mass.
   If the actual pressure is greater than 1 atm but assumed to be 1 atm, this infers a lesser number of moles and a higher molar mass.
   b. % error $= \dfrac{\text{pressure difference}}{\text{actual pressure}} \times 100$

3. % error $= \dfrac{\text{volume difference}}{\text{actual volume}} \times 100$

4. *High.* The mass of the liquid unknown (as well as the mass of the vapor) is also measured contributing to a higher mass and thus a higher molar mass.

5. *High.* A higher recorded temperature at a measured pressure and volume calculates to fewer moles of gas in the flask. A fewer number of moles infers a larger molar mass of the gas.

LABORATORY QUIZ
1. a. A 137-mL flask contains 0.481 g of a gas at 767 Torr and 22°C. What is the molar mass of the gas? [Answer: 84.2 g/mol]
   b. What is the density of the gas at STP? [Answer: 3.76 g/L]

2. If liquid unknown remains in the flask and the heating is stopped, is the calculated molar mass high or low?   [Answer: high]

3. A student misread the experiment and places 0.6 mL of unknown in the flask instead of 6 mL.
    a. What assumption in the experimental design is no longer valid?
                [Answer: the vapor may no longer displace all of the air from the flask]
    b. How will this affect the reported molar mass of the unknown? Explain.
        [Answer: the molar mass of the unknown will be reported less than actual if its molar mass is greater than the molar mass of air]

A Calcium Carbonate Mixture and the Molar Volume of Carbon Dioxide

INTRODUCTION	This experiment requires the application of Dalton's law of partial pressures, a combination of Boyle's and Charles' laws, and stoichiometry to determine the molar volume of CO_2 and the percent $CaCO_3$ in a mixture. Because of the many principles involved in this experiment, students should have already covered this material in lecture.
WORK ARRANGEMENT	Partners are suggested.
TIME REQUIREMENT	3 hours

LECTURE OUTLINE

1. Follow the Instruction Routine outlined in "To the Laboratory Instructor."

2. Review the stoichiometry of the reaction and the procedure by which each of the following are to be determined from the collected data:
   - the mass of CO_2 evolved
   - the volume of CO_2 evolved
   - the pressure of the CO_2
   - the temperature of the CO_2
   - the calculation of the molar volume of CO_2
   - the percent $CaCO_3$ in the initial reaction mixture.

3. In your lecture, note the significance of the Alka-Seltzer® or the acidified solution of baking soda, $NaHCO_3$. Because of the small solubility of CO_2 in water, the water in CO_2-collecting test tube is saturated with CO_2.

4. **Part A.2.** A calculation for the mass of $CaCO_3$ that is to be used for analysis is to be completed for the sample preparation. (A 50-mL volume of CO_2 at STP requires the complete reaction of about 0.22 g of $CaCO_3$)

5. Ideal gases have a molar volume of 22.4 L/mol. Students will think any other value is incorrect—point out that CO_2 is a *real* gas and therefore they should report their results "as is" and *not* try to hedge their data to make their results "come out right".

CAUTIONS & DISPOSAL

- Caution students on the insertion of the glass tubing through the rubber stopper (Technique 1)

- 3 *M* HCl is the only chemical that may be of any danger

TEACHING HINTS

1. **Part A.2.** The mass of the CO_2 generator and beaker (Figure 18.2) may exceed the capacity of your ±0.001 g balance. If so, use a ±0.01 g balance in its place (read Technique 6).

2. **Part A.3.** Watch the progress of the construction of the apparatus shown in Figure 18.3.

3. **Part A.4.** You may need to assist students in setting up the water-filled gas collecting test tube. Your approval of the experimental setup is required before students begin Part B; make sure that the stopper and tubing are air-tight and secure.

4. **Part B.** Gentle shaking is recommended in order to generate a slow evolution of CO_2 gas.

5. **Part C.2.** Transferring the gas-collecting 200-mm test tube to a leveling tank without spilling any water is challenging. Some students will become enraged with their inexperience and ineptness of the transfer—keep cool!

6. **Part C.3.** If the mass of the water-filled test tube and beaker exceeds the capacity of the balance, then volume measurements of the water in partially-filled and filled test tube must be used to measure the volume of CO_2 generated (see footnote 3).

7. Some assistance with the calculations may be necessary.

CHEMICALS REQUIRED	Alka-Seltzer®	1 tablet
	or $NaHCO_3$	5 g
	3 M HCl	20–25 mL

SUGGESTED UNKNOWNS

For the calculation in Part A.2, 0.22 g $CaCO_3$ generates 50 mL CO_2 at STP. Therefore the mass of $CaCO_3$ in the unknown mixture should not exceed 0.25 g in the sample. This mass produces about 56 mL of CO_2; the volume of a 200-mm test tube is about 75 mL.

Each student requires about 0.5 g of unknown.

Suggested mixtures by mass:
Sample I.	10 g $CaCO_3$	0.22 g generates ≈50 mL of CO_2
Sample II.	7.5 g $CaCO_3$, 2.5 g NaCl	0.22 g generates ≈37 mL of CO_2
Sample III.	5.0 g $CaCO_3$, 5.0 g NaCl	0.22 g generates ≈25 mL of CO_2
Sample IV.	2.5 g $CaCO_3$, 7.5 g NaCl	0.22 g generates ≈12 mL of CO_2

SPECIAL EQUIPMENT

weighing paper		2	gas delivery tube w/rubber stopper for	
balance (±0.001 g or ±0.01 g)		1	200-mm test tube	1
pneumatic trough		1	ring stand and test tube clamps	2
110°C thermometer		1	glass plate	1
laboratory barometer			leveling tank: 4-L beaker or sink	1
200-mm test tube		2	"Waste Mercury" container	

PRELABORATORY ASSIGNMENT

1.
   a. 729 Torr
   b. 0.0276 L
   c. 1.09×10^{-3} mol
   d. $27.6 \text{ mL} \times \dfrac{729 \text{ Torr}}{760 \text{ Torr}} \times \dfrac{273 \text{ K}}{293 \text{ K}} = 24.4$ mL or 0.0244 L;

   $\dfrac{0.0244 \text{ L}}{1.09 \times 10^{-3} \text{ mol}} = 22.4$ L/mol

2.
   a. $CaCO_3(s) + 2 H_3O^+(aq) \rightarrow Ca^{2+}(aq) + 3 H_2O(l) + CO_2(g)$
   b. 0.345 g CO_2; 7.84×10^{-3} mol CO_2
   c. 7.84×10^{-3} mol $CaCO_3$; 0.785 g $CaCO_3$
   d. 34.0 % $CaCO_3$

3.
   a. The water levels inside and outside the 200-mm gas-collecting test tube are adjusted to be the same as in a leveling tank. This equalizes the pressure of the "wet" CO_2 gas with atmospheric pressure, measured with a barometer.
   b. The mass of CO_2 is determined by subtracting the mass of the $CaCO_3$ mixture after heating from the mass of the same mixture before heating.

LABORATORY QUESTIONS

1.
   a. *Less.* The pressure in the test tube is less than atmospheric pressure because the pressure exerted by the "wet" CO_2 does not force the liquid level down to equal that of the pressure caused by the atmosphere.
   b. *Decrease.* To equilibrate the gas pressure, the test tube must be lowered; this increases the CO_2 pressure and therefore decreases its volume.
   c. *Greater.* The larger volume (lower pressure) would infer more moles of CO_2 in the collected volume.

2. a. *High.* The mass and moles of $CaCO_3$ are determined from the measured moles of CO_2 collected in the test tube and the stoichiometry of Equation 18.1. The greater the volume of the measured CO_2 the greater the reported number of moles of $CaCO_3$. The mass and moles of CO_2 would be unaffected by the air bubble.

    b. *High.* The additional volume (due to air) would increase the $\dfrac{volume}{mole}$ ratio.

LABORATORY QUIZ 1. A 23-mL volume of O_2, collected over water at 752 Torr and 22°C, is produced from the thermal decomposition of $KClO_3$. The vapor pressure of water at 22°C is 19.8 Torr.
$$2\, KClO_3(s) \rightarrow 2\, KCl(s) + 3\, O_2(g)$$
  a. How many moles of O_2 are collected?   [Answer: 9.18×10^{-4} mol]
  b. Calculate the mass of $KClO_3$ that decomposes.   [Answer: 75.0 mg]

2. What are the units for the molar volume of a gas?   [Answer: liters/mole]

3. What is the approximate volume of a 200-mm test tube?   [Answer: 75 mL]

4. a. State the purpose for adjusting the water level inside the 200-mm gas-collecting test tube equal to that of the leveling tank.
      [Answer: to adjust the pressure of the "wet" gas to that of atmospheric pressure]
  b. Does this affect the reported number of moles of CO_2 collected? Explain.
      [No. See answer to Laboratory Question 2a.

5. a. The density of helium gas at STP is 0.1785 g/L. Calculate its molar volume.
      [Answer: 22.43 L/mol]
  b. The density of SO_2 gas at STP is 2.927 g/L. What is its molar volume?
      [Answer: 21.89 g/L]

Experiment 19
Aluminum Analysis

INTRODUCTION	The chemical principles of this experiment are similar to those of Experiment 18; however, this experiment has an "applied" aspect to it. The ideal gas law, Dalton's law, and stoichiometry principles are used for the analysis.
WORK ARRANGEMENT	Partners. The construction of the apparatus and the collection of the $H_2(g)$ requires the presence of at least two chemists.
TIME REQUIREMENT	2.5 hours for two trials.

LECTURE OUTLINE	1. Follow the Instruction Routine outlined in "To the Laboratory Instructor."
	2. Students are to complete a calculation in Part A.1 before the beginning the Experimental Procedure. About 0.036 g of aluminum is required to generate 45 mL of $H_2(g)$ (at STP). If a 200-mm test tube (75 mL volume) is used as the $H_2(g)$ collector, the mass of aluminum can be increased to 0.056 g for the generation of 70 mL $H_2(g)$ at STP.
	3. A review of the calculations using Equations 19.2–19.5 is beneficial.
	4. Indicate which gas-collecting vessel is used in your laboratory, a gas buret or a 200-mm test tube. Footnotes 5 and 6 detail the use of the 200-mm test tube for gas collection.
	5. Note that in Figure 19.2, the HCl(aq) must *not* come into contact with the aluminum metal prior to closing the system.

CAUTIONS & DISPOSAL	• If the 6 M HCl comes into contact (Part A.2) with the skin, wash immediately.
	• **Part B.1.** The insertion of glass tubing through a one-hole stopper (Technique 1) is *always* dangerous if done incorrectly. Make sure you inform students of the proper procedure!
	• The contents of the reaction mixture, after reaction, are to be discarded into a "Waste Acids" container. The total wastes from the laboratory can be neutralized and discarded into the sink.

TEACHING HINTS	1. **Parts A.2 and B.1.** *First*, place the 6 M HCl in the 200-mm test tube; *second*, carefully lower the 75-mm test tube containing the aluminum sample into the HCl solution, *third*, stopper the 200-mm test tube with a one-hole stopper, fitted with a gas delivery tube—no other sequence is advisable.
	2. **Part B.2.** If the U-tube is not available, students should refer to Appendix A, cut a 15–20-cm piece of 6-mm glass tubing, bend it to a 45° angle, and fire polish the ends. Also refer to footnote 5 for a variation of the hydrogen collecting apparatus. The glass-tubing connections must be *tight*; wire may be used to tighten the connections.
	3. **Part B.3.** Inspect *and approve* the apparatus before students mix the HCl(aq) with the aluminum metal.
	4. **Part C.2.** The purpose of adjusting the water level in the gas-buret (or 200-mm test tube) to that of the leveling tank is to establish the pressure of the "wet" $H_2(g)$ equal to atmospheric pressure. See Figures 19.3 and 19.4.

| CHEMICALS REQUIRED | aluminum pieces | 0.1 g |
| | 6 M HCl | 30 mL |

| SUGGESTED UNKNOWNS | Aluminum metal from a soft drink can (each has a mass of about 17.5 g!!), aluminum TV tray, or aluminum foil 0.1 g |

SPECIAL EQUIPMENT	800-mL or 1-L beaker (or pneumatic trough)	1	rubber (or Tygon) tubing	25 cm
	weighing paper	2	110°C thermometer	1
	balance (±0.001 g)		laboratory barometer	
	gas delivery tube w/rubber stopper for 200-mm test tube	1	ring stand and buret clamp	
			ring stand and test tube clamp	
	leveling tank (4-L beaker or sink)		glass plates	2
	50-mL gas buret (or 200-mm test tube)	1	U-tube (Appendix A)	1
			"Waste Acids" container	
			"Waste Mercury" container	

PRELABORATORY ASSIGNMENT

1. a. 44.8 mL at STP
    b. 43.9 mL at STP

2. 73.2 %Al

LABORATORY QUESTIONS

1. *High.* The air contributes to a greater recorded volume and moles of $H_2(g)$ in the buret. According to Equation 19.1, the greater the calculated moles of $H_2(g)$, the greater is the reported percent aluminum in the sample.

2. *Less.* According to Equation 19.1, the fewer the moles of $H_2(g)$ collected, the lower is the reported percent aluminum in the sample.

3. *Low.* A higher assumed temperature infers a higher vapor pressure of water and a lower pressure (and fewer moles) of $H_2(g)$ (see Equation 19.3). From $PV = nRT$, a larger assumed temperature also calculates to a fewer moles of $H_2(g)$.

4. *High.* If the combined pressure of the $H_2(g)$ and water vapor in the buret is assumed to be at atmospheric pressure, then a greater than actual pressure (and moles of) $H_2(g)$ will be assumed.

LABORATORY QUIZ

1. A 42.3-mL volume of $H_2(g)$ is collected at 20°C and 748 Torr from the reaction of a 0.0361-g sample of aluminum with excess hydrochloric acid. Calculate the percent aluminum in the sample. The vapor pressure of water at 20°C is 17.5 Torr.
    $2 Al(s) + 6 HCl(aq) \rightarrow 2 AlCl_3(aq) + 3 H_2(g)$ [Answer: 84.3% Al]

2. If the pressure of the collected "wet" $H_2(g)$ in the gas buret is not corrected for the vapor pressure of water, is the reported mass of aluminum high or low? Explain.
    [Answer: high]

Molar Mass of a Solid

INTRODUCTION	The molar mass of a compound is discussed very early in the lecture, but the methods for measuring molar mass are not. The procedure for determining the molar mass of a gas described in Experiment 17 uses gas relationships; in this experiment the molar mass of a solid is determined by measuring the lowering of the freezing point (colligative property) of a solvent from the formation of a solution.
WORK ARRANGEMENT	Partners.
TIME REQUIREMENT	3 hours for the measurement of the freezing point of the solvent and the three freezing points of the solutions. The analysis of plotted is required.

LECTURE OUTLINE

1. Follow the Instruction Routine outlined in "To the Laboratory Instructor."

2. Review the principle for determining the molar mass of a compound using the freezing point lowering technique. Use Figures 20.2 and 20.3 in your discussion.

3. Inform students that colligative properties are dependent only on the moles of solute particles and *not* on the nature of the solute particles.

4. In this experiment, we assume no dissociation of the solute.

5. A clean, dry 200-mm test tube needs to be available at the beginning of the laboratory period. Time is wasted in waiting for a test tube to dry. Students can either clean the test tube during the preceding laboratory period and then allow it to air-dry in their lab drawer or clean it before your lab lecture and place it in the drying oven.

6. Note that additional solute is added to the solution in Parts B.4 and B.5; the preparation of a new solution for each trial is *not* required.

CAUTIONS & DISPOSAL

• Thermometers are not stirring rods!

• If a thermometer is broken, collect as much of the mercury as possible and then dust the area with powdered sulfur.

• Place the mercury in a "Waste Mercury" bottle.

• Discard the waste cyclohexane solutions in a "Waste Organic Liquids" container.

TEACHING HINTS

1. *t*-Butanol is also an effective solvent for the determination of the molar mass of a solid.

2. **Parts A.3.** The freezing point of cyclohexane is 6.5°C. Continuous stirring of the cyclohexane solvent is necessary to avoid supercooling and uneven cooling. Approve the student graphs according to the guidelines in Appendix C.

3. **Part A.4.** Some assistance may be necessary for plotting the temperature/time curves and determining T_f. Be certain students correctly obtain T_f from the plotted data.

4. **Part B.1.** All of the solute (solid or liquid) must dissolve—no solute should adhere to the test tube wall.

5. **Part B.2.** Stir the solution continuously during the cooling; most errors result from non-uniform cooling, supercooling, and a rapid cooling rate. The stirring should not be so rapid that splashing of the solution occurs. Minimize the supercooling effect.

6. **Parts B.4 and B.5.** The additional solute proportionally decreases the freezing point of the solution. Remind students that the mass of the solute in solution is a *sum* of the masses of solute from Parts B.1, B.4 (Trial 2), and B.5 (Trial 3).

7. **Part B.6.** Be certain students correctly determine T_f for each solution and obtain ΔT_f from the plotted data. Approve all student graphs. Insist on good graphs; see Appendix C.

8. A molar mass of ±5% is a good result in this experiment.

CHEMICALS REQUIRED		
cyclohexane		15 mL
t-butanol (optional)		15 mL

SUGGESTED UNKNOWNS

Approximately 1 g of sample is required per student for the unknown. The suggested mass of the sample in Column 3 is based upon a $\Delta T \approx 2°$ for 0.010 kg of cyclohexane.

Compound	Formula	Mass of Sample (B.1)	Additional Mass of Sample (B.4 plus B.5)	Molar Mass (g/mol)
anthracene	$C_{14}H_{10}$	0.18 g	+ 0.4 g	178.23
naphthalene	$C_{10}H_8$	0.13 g	+ 0.3 g	128.19
benzophenone	$C_6H_5COC_6H_5$	0.18 g	+ 0.4 g	182.22
biphenyl	$C_6H_5-C_6H_5$	0.15 g	+ 0.3 g	154.21
camphor	$C_{10}H_{16}O$	0.15 g	+ 0.3 g	152.23
p-dibromobenzene	$1,4\text{-}Br_2C_6H_4$	0.24 g	+ 0.5 g	235.91
p-dichlorobenzene	$1,4\text{-}Cl_2C_6H_4$	0.15 g	+ 0.3 g	147.00
diphenylamine	$(C_6H_5)_2NH$	0.17 g	+ 0.4 g	169.23
p-nitrotoluene	$4\text{-}NO_2C_6H_4CH_3$	0.14 g	+ 0.3 g	137.13
sulfur	S_8	0.26 g	+ 0.5 g	256.48
toluene	$C_6H_5CH_3$	0.10 g	+ 0.2 g	92.15

SPECIAL EQUIPMENT

ice			600-mL beaker	1
200-mm test tube	1		ring stand and test tube clamp	1
balance (±0.01 g)			weighing paper	3
wire stirrer	30-40 cm		"Waste Organic Liquids" container	
110°C thermometer	1		"Waste Mercury" bottle	
timer with a second hand				

PRELABORATORY ASSIGNMENT

1. 153 g/mol

2. 81.2°C

3. Pure solvents, as all pure substances, have a fixed freezing point. The freezing of a solution is not constant; the freezing temperature of a solution continues to decrease, however, because as the solvent freezes the remaining solution becomes more concentrated, resulting in a lowering of the freezing temperature.

4. a. Undercooling: the temperature falls below the normal freezing point without freezing the cyclohexane (or solution).
   b. Since undercooling is an unstable condition, continuous stirring of the cyclohexane (or solution) minimizes this phenomenon.

5. *Student 2.* The freezing point of the solvent will be the same for each student, but because the solution is more concentrated for student 2, a greater freezing point change will be observed.

LABORATORY QUESTIONS

1. *Low.* From $\Delta T_f = k_f m$, a larger ΔT_f implies more moles of solute producing a smaller molar mass (g/mol).

2. a. ΔT_f *decreases.* Less solute particles decreases the freezing point change.
   b. ΔT_f *decreases.* Less solute particles means a smaller freezing point change.

3. *No effect.* Since only a temperature *change* (in °C) is measured in the experiment, the miscalibration has no effect on the colligative property measurement or the moles of solute.

4. *Less.* With less solute dissolved in the cyclohexane, the ΔT_f is also less ($\Delta T_f = k_f m$).

5. *No effect.* The added moles of solute proportionally changes the freezing point of the given solvent system. (Experimentally however, the freezing point of the pure solvent could not be measured).

LABORATORY QUIZ

1. Define a colligative property of a solution.[Answer: a property of a liquid that depends only upon the number of solute particles present in the mixture]

2. List three colligative properties of solutions.[Answer: lowering of the vapor pressure, the lowering of the freezing point, the increase in the boiling point, and osmotic pressure]

3. Which solution has the lowest freezing point?
   a. 0.10 mol NaCl in 1 kg of water
   b. 0.10 mol glucose in 1 kg of water
   c. 0.10 mol $CaCl_2$ in 1 kg of water
   d. 0.10 mol KNO_3 in 1 kg of water
   Justify your choice.   [Answer: c; more moles of solute particles]

4. A 0.947-g sample of an unknown solute dissolved in 15.0 g of *t*-butanol produces a solution with a freezing point of 22.7°C. What is the molar mass of the solute? For *t*-butanol, $T_f = 25.5$°C, $k_f = 9.1$°C·kg/mol.   [Answer: 205 g/mol]

5. Explain the likely appearance of a temperature-time curve at the freezing point of a solvent if *no* stirring is conducted.   [Answer: supercooling at the freezing point]

6. From the following data, plot the temperature-time curve, and estimate the freezing point of the solution.

Time	Temperature	Time	Temperature
10	25.7	70	21.4
20	24.8	80	21.3
30	23.8	90	21.4
40	22.7	100	20.9
50	22.0	110	20.0
60	21.6	120	19.1

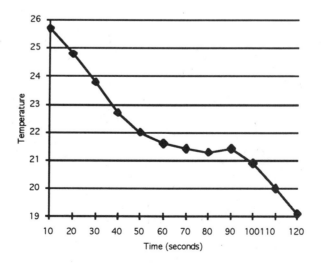

*7. The freezing point of a 1.0 m HX aqueous solution is –2.13°C. k_f for water is
1.86 °C•kg/mol
    a.   Calculate the moles of solutes particles present in 1 kg of water.[Answer: 1.15 mol]
    b.   What is the percent dissociation of HX?   [Answer: 14.5%]

Experiment **21**
Calorimetry

INTRODUCTION	Energy (heat) is evolved or absorbed in many chemical and physical processes. In this experiment students measure heat changes with a calorimeter that consists of two polystyrene coffee cups and a thermometer; the results are generally within ± 5% of the literature values.
WORK ARRANGEMENT	Partners. Students collect time/temperature data. The data gathering is facilitated with the shared experiences.
TIME REQUIREMENT	This is a long experiment; you may select only parts of Experimental Procedure for student to complete About 3 hours for collecting the experimental data for Parts A, B, and C and 1 hour for the calculations and graphing analysis are required for the average (or ill-prepared) students. Inform students which parts of the experiment are to be completed (and omitted, if any).

LECTURE OUTLINE

1. Follow the Instruction Routine outlined in "To the Laboratory Instructor."

2. Energy changes accompany all chemical and physical changes. Exothermic (heat loss) changes are assigned a negative enthalpy value; endothermic (heat absorbed) changes are positive. The directional signs of heat flow is sometimes confusing; it is advisable to be well prepared on "signs" before explaining any part of this experiment to the students.

3. Review the purpose and method of measurement for each part of the Experimental Procedure that is assigned for completion.

4. Temperatures are to be approximated to ±0.21°C and water volumes to ±0.2 mL. Sloppiness significantly affects the results.

5. Rinse the thermometer and calorimeter thoroughly after each measurement.

6. Discuss the time (abscissa)–temperature (ordinate) plot, Figure 21.5. Students may require assistance in setting up a properly labeled graph with appropriately expanded scales (Appendix C). Graph paper is provided at the back of the manual.

7. Stir all mixtures quickly and thoroughly. However, rapid stirring of the solution in the calorimeter may cause some heating or cooling from aeration—inform students to be aware of all possible errors.

CAUTIONS & DISPOSAL

•   Thermometers are not stirring rods!

•   Collect spilled mercury as best possible, then dust the area with powdered sulfur.

•   Place the mercury from broken thermometers in a "Waste Mercury" bottle.

•   The metals used in Part A are to be returned to appropriately labeled containers. All other test samples may discarded in the sink, followed by a generous flush with tap water.

TEACHING HINTS

1. **Part A.1.** No water is in the 200-mm test tube. The water level *in the beaker* is to be above the metal, but the metal remains dry.

2. **Part A.4.** Don't splash water from the calorimeter when the hot metal is transferred from the test tube to the calorimeter. Swirl to stir the water to obtain thermal equilibrium in the calorimeter. The thermometer is *not* a stirring rod. The temperature and recorded time intervals are to be tabulated.

3. **Parts A.5, B.4, C.4.** Approve the plotted data. The temperature-time graphs should be constructed according to the graphing guidelines in Appendix C. Generally the decrease in temperature after the maximum recorded temperature will not be as dramatic as what may be implied from Figure 21.5.

4. **Part B.1.** The 50 mL of 1.1 M HCl produces an excess of H_3O^+ to ensure neutralization of all of the OH^-. The calculations are based on the 0.050 mol OH^-, the limiting reactant, in the system.

5. **Part B.6.** Substitution of 1.1 M HNO_3 for 1.1 M HCl produces the same heat (enthalpy) of neutralization, $\Delta H_n = -57.8$ kJ/mol H_2O or -13.8 kcal/mol H_2O. Remember 1 mol OH^- produces 1 mol H_2O.

6. **Part C.3.** The solution is stirred until the salt dissolves. Note that with most salts the dissolving process is endothermic and the temperature decreases; therefore, the plot of temperature versus time will be inverted from what appears in Figure 21.5.

CHEMICALS REQUIRED		
1.1 M HCl	110 mL	
1.1 M HNO_3	110 mL	
1.0 M NaOH (standardized)	200 mL	

SUGGESTED UNKNOWNS

Part A. The mass of metal at 100°C that produces $\Delta T \approx 3$°C for 20 g of water at 20°C is given in the right column.

Metal	Molar Mass (g/mol)	Specific Heat Capacity (J/(g•°C))	Mass of Metal for $\Delta T \approx 3$°C
copper	63.5	0.385	8.5
lead	207	0.130	25
aluminum	27.0	0.900	3.6
zinc	65.4	0.389	8.4
tin	119	0.218	14.9
chromium	52.0	0.469	6.9
manganese	54.9	0.481	6.8
nickel	58.7	0.444	7.3
iron	55.8	0.46	7.1

Part C. The mass of *anhydrous* salt that produces $\Delta T \approx 5$°C in 20 g of water is given in the right column.

Salts (10 g)	Molar Mass (g/mol)	ΔH_s* (kJ/mol)	Specific Heat* (J/(g•°C))	Mass of Salt for $\Delta T \approx 5$°C
NH_4Cl	53.49	+ 14.8	1.57	1.6
NH_4NO_3	80.08	+ 25.7	1.74	1.3
$(NH_4)_2SO_4$	132.14	8.778[b]	1.409[c]	7
KOH	56.11	−57.6	1.16	0.4
KNO_3	101.11	+ 34.9	0.95	1.2
KCl	74.55	+ 17.2	0.688	1.8
KBr	119.01	+19.9	0.439	2.5
KI	166.01	+ 20.3	0.319	3.5
K_2SO_4	174.27	+ 23.8[a]	0.754	3.1
NaOH	40.00	− 44.5	1.49	0.38
NaCl	58.44	3.88	0.864	6.7
NaI	149.89	− 7.53	0.348	8.6
Na_2SO_4	142.04	− 2.4[a]	0.903	34
Na_2CO_3	105.99	− 26.7[a]	1.06	1.7
$NaNO_3$	84.99	+ 20.4	1.09	1.8

$Na_2S_2O_3$	158.11	35.69[b]	1.45[c]	1.9
LiCl	42.39	− 37.0	1.13	0.48
LiOH	23.95	− 23.6	2.07	0.43
$Ca(OH)_2$	74.09	− 16.7[a]	1.18	1.9

*Data was obtained from *Handbook of Chemistry and Physics*, 66th Edition. Conversions were made from calories to joules and from moles to grams wherever appropriate.
[a]Atkins and Beran, *General Chemistry, 2nd Edition, Updated Version*, Scientific American Books, New York, 1992, p. 422.
[b]Landolt-Bornstein, New Series
[c]International Critical Tables

SPECIAL EQUIPMENT	wire stirrer balance (±0.01 g) balance (±0.001 g) polystyrene coffee cups (6–8 oz w/lid) thermometer, 110°C and clamp weighing paper	 2 1 3	ring stand and iron support rings ring stand and test tube clamp boiling chips Bunsen burner "Waste Mercury" container labeled metal solids containers

PRELABORATORY ASSIGNMENT

1. 0.512 J/(g•°C); ≈ 49 g/mol

2. *Less*. The glass beaker is a better conductor of heat resulting a heat loss from the system that would be greater than the heat loss from a styrofoam calorimeter.

3. The excess moles of H_3O^+ ensures the complete reaction of the 0.050 mol OH^-; the OH^- is the limiting reactant and thus limits the heat evolved in the reaction.

4. As heat is generated in the system, either from the transfer of a warmer metal (Part A) or from the acid-base reaction in Part B, the calorimeter is constantly absorbing some of the energy, thus the maximum temperature is never measured. By extrapolation, the maximum temperature can be interpreted.

5. a. Lattice energy: the energy required to vaporize one mole of salt into its gaseous ions.
   b. Hydration energy: the energy released when gaseous ions of a salt is hydrated with water molecules.

LABORATORY QUESTIONS

*1. If a known mass of warm water at a measured temperature is added to a known (and nearly equal) mass of cool water also at a measured temperature in the calorimeter, a measured final temperature is obtained. If the heat loss of the warm water is greater than the heat gained by the cool water, then the heat loss is to the calorimeter. The heat loss over the temperature change of the cool water in the calorimeter is the **heat capacity** of the calorimeter, also called the **calorimeter constant** (J/°C). The calorimeter constant can then can be included in subsequent calculations involving heat changes in the calorimeter.

2. *Lower*. Since the maximum recorded temperature is slightly less than the extrapolated temperature, the ΔT_M for the metal is recorded too large and the ΔT_{H_2O} is recorded too small. The larger ΔT_M for the metal results in a smaller specific heat capacity. See Equation 21.4.

3. All strong acids and bases are 100% ionized; therefore, all reactions involve only the reaction of H_3O^+ with OH^-: $H_3O^+(aq) + OH^-(aq) \rightarrow 2 H_2O(l)$

4. *No effect*. The temperature measurement for ΔH_n (as are all temperature measurements in this experiment) is a temperature *difference*, not an absolute temperature. The 4°C miscalibration would not appear in the calculations.

5. 147 J/g KBr; 17.5 kJ/mol KBr

LABORATORY QUIZ   1.   A 24.7-g sample of an unknown metal, heated to 98.4°C , is transferred to 25.0 mL of H_2O at 21.4°C. The equilibrium temperature is 27.3°C.
  a.   What is the specific heat capacity of the metal?   [Answer: 0.351 J/(g•°C)]
  b.   What is the approximate molar mass of the metal?   [Answer: 71 g/mol]

2.   Why is the molar ΔH_n (enthalpy of neutralization) of a strong acid and a strong base nearly constant?[Answer: the net ionic reaction is $H_3O^+(aq)$ + $OH^-(aq)$ → 2 $H_2O(l)$ for all strong acid-strong base reactions]

3.   Identify the two theoretical factors used to explain why the dissolving of a salt may be an exothermic or endothermic process.
                           [Answer: the lattice energy and the hydration energy]

4.   When 25.0 g of H_2O at 80.4°C is added to 100 g H_2O at 24.7°C in a calorimeter (also at 24.7°C), the equilibrium temperature is 35.0°C.
  a.   How much heat is lost to the calorimeter?   [Answer: 4744 J]
  b.   Calculate the heat capacity (J/°C) of the calorimeter.   [Answer: 42.6 J/°C]

Factors Affecting Reaction Rates

INTRODUCTION	A semi-quantitative study of factors affecting reaction rates complements the discussion of kinetics in class. Students observe and measure the reaction rates for several reactions and, as a result, derive their own conclusions about factors affecting reaction rates. Predictions of rates are derived from plotted the semi-quantitative data. The completion of this experiment requires you, as a lab instructor, to be prepared. Do *not* plan to give a quiz during this period.
WORK ARRANGEMENT	Partners. A discussion of the observations and a collaboration of the analysis of the data are important learning experiences.
TIME REQUIREMENT	3.5 hours, *if* the students are well-prepared. Time is required for the analysis of the data; calculations are required before some data is plotted. To reduce time for the experiment, Parts B *or* C and Parts E *or* F can be omitted. Much of the time is used for setting up each apparatus and for graphing/interpreting data.

LECTURE OUTLINE

1. Follow the Instruction Routine outlined in "To the Laboratory Instructor."

2. Review the five factors affecting reaction rates; four of which are addressed in the Experimental Procedure—the surface effects of the reaction rates is excluded.

3. Identify which parts of the experiment are to be completed/omitted.

4. Discuss the use of the 24-well trays for Parts A, D, E, and F of the experiment—less chemicals are used, but the observations are just as informative as when test tubes are used.

5. Divide the class into three groups to avoid congestion at the reagent table. Each group should complete the experiment as follows:
   Group I.   A, B, C, D, E, F
   Group II.   C, D, E, F, A, B
   Group III.   E, F, A, B, C, D

6. **Parts B and/or C.** Advanced students can be assigned the determination of the activation energy for the reaction. A plot of ln k vs. $^1/_T$ yields a slope equal to $^{E_a}/_R$. The ln k is proportional to the reaction rate.

CAUTIONS & DISPOSAL

- The strong acids used in Part A should be handled carefully. Take the appropriate precautions to prevent acid spills and to clean up acid spills.

- Properly dispose of all chemicals and solutions used in the experiment.

- A disposal container marked "Waste Inorganic Test Solutions" should be available in the laboratory. None of the chemicals used in the experiment are considered dangerous to the environment if diluted.

TEACHING HINTS

1. Students tend to anticipate experimental observations—be inquisitive during the entire laboratory period.

2. The setup of a hot water bath should be the first procedure in the laboratory.

3. **Part A.1.** The reaction rates are nearly the same for 3 M H_2SO_4 and 6 M HCl, only slightly slower for 6 M H_3PO_4, and decidedly slower for 6 M CH_3COOH. Remember that the size of the magnesium strip and the extent to which it is polished are also factors in observing the reaction rates.

4. **Part A.2.** The reaction rate of 6 M HCl with Zn is rapid, Mg is very rapid, but Cu does not react.

5. **Parts B and C.** The time required for the change to appear decreases with increasing temperature. Note that while the time decreases, the reaction rate, $\left(\dfrac{\Delta \text{ color change}}{\Delta t}\right)$, increases. The reaction rate also depends on slight concentration variations in the reaction mixtures.

   A major source of error is contamination. Students should rinse the pipets with the reagents before dispensing a measured aliquot; pipets are not to be mixed from one reagent to the other.

   Students may require some assistance in the plots of temperature versus reaction time. See Appendix C. Approve the student graphs; the graphs are to be constructed so that interpretations and extrapolations can be easily made.

6. **Part D.** The presence of the MnO_2 catalyst significantly increases the decomposition rate of the hydrogen peroxide solution.

7. **Part E.1.** As an extra assignment, our students are asked to prepare the 3 M, 1 M, and 0.1 M HCl solution from the 6 M HCl stock solution.

8. **Part E.2.** The rate of the disappearance of magnesium increases with an increase in the HCl concentration, varying from about 7 seconds for 6 M HCl to about 80 seconds for 0.1 M HCl. Keep in mind that, while quantitative measurements are made, the data will vary from student to student. Approve the student graphs.

9. **Part F.** Students are to calibrate the Beral (or dropping) pipets—the volume per drop is determined by counting the number of drops in one milliliter, dispensed into a 10-mL graduated cylinder.

   The 24-well tray is ideal for observing the data of this reaction system. The reaction rate increases (time decreases) with successive larger volumes of 0.01 M HIO_3. Our data are (rounded to the nearest second): Well A1: 4:05 min; Well A2: 1:22 min; Well A3: 43 s; Well A4: 31 s; Well A5: 22 s—again, several factors affect these times and should not be interpreted as *accurate* times. Students again may require assistance in the graphing of the data. Approve the graphs.

CHEMICALS REQUIRED	Mg, Zn, Cu strips		3 % H_2O_2	2 mL
	6 M HCl	2 mL	$MnO_2(s)$	"pinch"
	6 M HNO_3	2 mL	Mg strip	2 cm
	6 M H_3PO_4	2 mL	6 M HCl	1 mL
	3 M H_2SO_4	2 mL	3 M HCl	1 mL
	0.1 M $Na_2S_2O_3$	6 mL	1 M HCl	1 mL
	0.1 M HCl	6 mL	0.1 M HCl	1 mL
	0.01 M $KMnO_4$/3 M H_2SO_4	3 mL	0.01 M HIO_3	3 mL
	3 M H_2SO_4	12 mL	starch	1 mL
	0.33 M $H_2C_2O_4$	15 mL	0.01 M H_2SO_3	5 mL

SPECIAL EQUIPMENT	24-well tray and Beral pipets	1 + 6	timer (in seconds)	
	steel wool		ring stand and iron support rings	
	110°C thermometer	1	test tube clamps	
	2-mL pipet and pipet bulb	2	Bunsen burner	
	ice and salt		balance ($\pm$0.001 g)	
	5-mL (0.1-mL graduated) pipets	4	"Waste Inorganic Test Solutions" container	

PRELABORATORY ASSIGNMENT	1.	a.	nature of reactants
		b.	temperature of the reaction
		c.	presence of a catalyst
		d.	concentration of reactants

2. a. enzymes catalyze biochemical reactions
    b. the concentration of pollutants is greater in the smog-laden air
    c. higher temperatures accelerate the spoilage
    d. gold and silver, by nature, corrode at a much slower rate
    e. higher temperatures accelerate reaction rates

3. Rate increases by a factor of 16.

4. a. $[HIO_3]_o$
    b. A Dependence of the HIO_3 Concentration versus Time for a Reaction

5. Part B. Time required for the appearance of the cloudiness due to elemental sulfur
    Part C. Time required for the disappearance of the purple MnO_4^- ion
    Part E. Time required for the disappearance of the magnesium metal
    Part F. Time required for the appearance of the deep blue $I_2 \cdot$ starch complex

LABORATORY QUIZ 1. Explain why the rate of disappearance of magnesium metal increases when added to 0.10 M HCl, 1.0 M HCl, and 6.0 M HCl solutions respectively.
[Answer: concentration factor]

2. Account for the rapid burning of wood sticks compared to that of wood logs.
[Answer: surface area of reactants]

3. Milk sours more quickly at room temperature than in a refrigerator. Explain.
[Answer: temperature factor]

4. Magnesium metal oxidizes in air more rapidly than silver metal. Explain.
[Answer: nature of the reactants]

5. Many biochemical reaction occur in the body with enzymes being present; however, the same reactants without the enzymes in a test tube react very slowly or remain unreacted. Characterize the function of the enzyme.
[Answer: presence of a catalyst]

6. Coal (assume it to be pure carbon) burns. List various methods that you would use to increase its rate of combustion.

Determination of a Rate Law

INTRODUCTION	Experiment 22 addressed the qualitative study of factors affecting reaction rates; Experiment 23 is a quantitative analysis of data used to determine the rate law for the reaction of KI with H_2O_2. Satisfactory data can be also collected when potassium persulfate, $K_2S_2O_8$, is substituted for the hydrogen peroxide in the experiment.

Students generally enjoy the data collection for this experiment, but have some difficulty with the its analysis. Advise students to complete the Prelaboratory Assignment and to carefully read the Experimental Procedure in advance.

This experiment uses a 6-well tray or 5–6 small (15–20 mL) beakers for the apparatus.

WORK ARRANGEMENT	Partners are needed for the collection of the data, for the mixing and the timing of the reaction mixtures.

TIME REQUIREMENT	1.5 hours for collecting the data, 1.5 hours for the calculations and the construction and interpretation of the graphs.

LECTURE OUTLINE	1. Follow the Instruction Routine outlined in "To the Laboratory Instructor."

2. Explain, in detail, the reaction between H_2O_2 and I^- and the monitoring of the reaction. It is not always clear to the student "why" the solution turns blue after a period of time and "why" the time required for the appearance of the blue color varies.

3. Emphasize that the reaction continues after the appearance of the blue $I_2 \bullet$ starch complex; its appearance only signifies the depletion of a quantitative amount of $S_2O_3{}^{2-}$ added to the solution. Also its depletion means a stoichiometric amount of I_2 has been generated (see Equations 23.2 and 23.9).

4. Encourage students to prepare additional test solutions (Table 23.1) so that additional data points can be plotted on their graphs.

5. Assist the students with their calculations, or take a representative set of data and follow through the steps on the Report Sheet.

6. Class data for the specific rate constant are to be collected for this experiment and a standard deviation of the class data is to be calculated. Advise students of the procedure for collecting the class data and review, if necessary, the calculation of the standard deviation.

CAUTION & DISPOSAL	• Provide a "Waste Iodide Salts" container for the disposal of the test solutions.

TEACHING HINTS	1. **Part A.** Clean glassware (or well plate) is of paramount importance in this experiment; H_2O_2 readily decomposes in the presence of foreign particles.

2. Pipets or burets must be used for preparing each solution, especially the 0.3 *M* KI, 0.2 *M* $Na_2S_2O_3$, and 0.1 *M* H_2O_2 solutions. Whereas the total volume does affect the reaction, the measured volumes of distilled water and the buffer and starch solutions are not as critical.

3. Encourage students to prepare additional reaction mixtures or to repeat some of the kinetic trials for more data points and for more precise data. Again, advise them of the importance of clean glassware.

4. **Part B.** Sample calculations for Kinetic Trial 1 are:

$$\text{mol } S_2O_3^{2-} = 1.0 \times 10^{-3} \text{ L} \times 0.020 \text{ mol/L} = 2.0 \times 10^{-5} \text{ mol } S_2O_3^{2-} \text{ consumed}$$

$$\text{mol } I_2 = 2.0 \times 10^{-5} \text{ mol } S_2O_3^{2-} \times \frac{1 \text{ mol } I_2}{2 \text{ mol } S_2O_3^{2-}} = 1.0 \times 10^{-5} \text{ mol } I_2 \text{ produced}$$

$$\frac{\Delta(\text{mol } I_2)}{\Delta t} = 1.0 \times 10^{-5} \text{ mol } I_2/\Delta t; \Delta t \text{ it measured in the reaction}$$

$$[I^-]_o = \frac{1.0 \times 10^{-3} \text{ L} \times 0.30 \text{ mol/L}}{1.0 \times 10^{-2} \text{ L}} = 3.0 \times 10^{-2} \text{ mol/L}$$

$$\log [I^-]_o = -1.52$$

$$[H_2O_2]_o = \frac{3.0 \times 10^{-3} \text{ L} \times 0.10 \text{ mol/L}}{1.0 \times 10^{-2} \text{ L}} = 3.0 \times 10^{-2} \text{ mol/L}$$

$$\log [H_2O_2]_o = -1.52$$

5. **Part C.** Two graphs are to be constructed—one to determine p, the order of the reaction with respect to $[I^-]$ and q, the order of the reaction with respect to $[H_2O_2]$. The slope of the line is the order of the reaction with respect to the respective reactant.

The log(rate), $\log [I^-]_o$, and $\log [H_2O_2]_o$ are all negative quantities. The plotting of only negative numbers on a graph (all data appears in the quadrant III) causes difficulty for many students in setting up a graph. The slope of the line is positive and the slope is calculated from $\frac{y_2 - y_1}{x_2 - x_1}$ where (x_2, y_2) and (x_1, y_1) are points on the line.

Students are sometimes insecure about drawing the best straight line when the data points do not lie in a straight line (and they usually do not); some students may need your assistance. Remember that the slope is determined from the line, not from the data points. Discourage students from obtaining the slope of the line with a least squares plot obtained from a calculator—the presence of a displaced data point is not recognized on a calculator.

A spurious data point is also reason for students to repeat some of the trials (Table 23.1) or to create other mixtures for additional data points.

7. **Part D.** Students may need help in determining a standard deviation for k', not only for their own data but also for their class data.

8. Approve all graphs.

CHEMICALS REQUIRED		
0.5 M CH$_3$COOH/0.5 M NaCH$_3$CO$_2$ buffer solution		10 mL
1% starch solution		2 mL
0.1 M H$_2$O$_2$ (or 0.1 M K$_2$S$_2$O$_8$, to two significant figures)		25 mL
0.3 M KI (to two significant figures)		10 mL
0.02 M Na$_2$S$_2$O$_3$ (stable for about 1 week)		8 mL

SPECIAL EQUIPMENT				
6-well plate or 5–6 beakers (10–20 mL)	1	10-mL grad. pipets for 0.1 M H$_2$O$_2$	1	
1-mL pipet and pipet bulb for buffer	1	110°C thermometer		
5-mL grad. pipets for 0.3 M KI and water	2	timer (in seconds)		
dropping pipet for starch	1	burets (optional) for dispensing some solutions		
1-mL pipet for 0.02 M Na$_2$S$_2$O$_3$	1	"Waste Iodide Salts" container		

PRELABORATORY ASSIGNMENT

1. a. The starch is an indicator in the reaction—its presence, regardless of its amount, combines with free I_2 in the reaction. The sodium thiosulfate reacts with I_2 as it is being generated. Its amount reflects the amount of I_2 that is generated during that time interval for the reaction.

   b. The starch intensifies the appearance of free I_2 generated in the reaction (Equation 23.10).

2. a. The sodium thiosulfate prevents the buildup of I_2 that is generated during the course of the reaction (Equation 23.9). The removal of the I_2 as it is generated reduces the probability of any back-reaction (of Equation 23.2) and thus maintains a constant reaction rate. The depletion of the $S_2O_3^{2-}$ leaves I_2 in the system which immediately complexes with the starch.

    b. $1.0 \text{ mL} \times \dfrac{0.02 \text{ mmol}}{\text{mL}} = 0.02$ mmol or 2×10^{-5} mol $S_2O_3^{2-}$

3. I⁻ reacts with H_2O_2 producing I_2 and H_2O (Equation 23.2). A measured amount of $Na_2S_2O_3$ consumes the I_2 as it is being produced. When the $Na_2S_2O_3$ is depleted, the timing of the reaction stops because the I_2 now reacts with starch forming the deep-blue I_2•starch complex, the signal to stop timing.

4. a. $0.88\ M\ H_2O_2$
    b. Dilute 11.4 mL of 3% H_2O_2 to 100 mL with water.
    c. Dissolve 4.98 g KI in 100 mL of solution.

LABORATORY QUESTIONS

1. a. *No effect.* Increasing the buffering capacity of the system has no effect on the concentration of reactants, and thus, on the reaction rate.
    b. *Increase.* According to the rate law, the greater the H_2O_2 concentration, the greater the reaction rate.
    c. *Increase.* See the rate law, Equation 23.4.
    d. *No effect.* If no dilution occurs and if the $Na_2S_2O_3$ concentration is not increased beyond the concentrations of the H_2O_2 *and* I⁻ , then the rate is unaffected; the time for the color change to occur is extended however.
    e. *No effect.* The intensity of the blue I_2•starch complex would increase with the higher concentration of the starch solution.

2. a. *Three.*
    b. Additional test reactions produce additional data points for the construction of the graph, increasing the overall reliability of the data and its interpretation.

3. The blue I_2•starch complex would appear immediately. The absence of $Na_2S_2O_3$ would not be present to delay the appearance of free I_2 in the solution; the pink color of the I_2 would slowly intensify as its concentration builds.
    b. When the $S_2O_3^{2-}$ is consumed in the reaction, free I_2 would be produced to form a pink solution.

4. a. *Decrease.* A lower concentration of I⁻ would decrease the reaction rate.
    b. *No effect.* The order of a reaction is unaffected by changes in concentration.
    c. *No effect.* A reaction rate constant is unaffected by changes in concentration.

LABORATORY QUIZ

1. The reaction of A with B producing C has the rate law:

    rate $\left(\dfrac{\Delta A}{\Delta t}\right) = k\ [A]^2[B]^0$

    Rearrange the rate law so that it represents an equation for a straight line.

    {Answer: $\log(\text{rate}) = \log k + 2 \log [A]$}

2. In the reaction of H_2O_2 and I⁻ , a quantitative amount of $S_2O_3^{2-}$ is added; starch solution is also added. The balanced equation for the reaction is:
    $H_2O_2(aq) + 2\ I^-(aq) + 2\ H^+(aq) \rightarrow I_2(aq) + H_2O(l)$
    a. What is the purpose of $S_2O_3^{2-}$?   [Answer: to reduce I_2 as it is generated]
    b. What is the purpose of the starch solution?
                               [Answer: to form a deep-blue complex with I_2]
    c. Why is it necessary that the $S_2O_3^{2-}$ be measured quantitatively, but *not* the starch?
              [Answer: to determine *quantitatively* the amount of I_2 that is generated]

3. A reaction occurs between $NH_4^+(aq)$ and $NO_2^-(aq)$ forming $N_2(g)$ and $H_2O(l)$.
    a. Write a balanced equation for the reaction.

[Answer: $NH_4^+(aq) + NO_2^-(aq) \rightarrow N_2(g) + 2 H_2O(l)$]

   b.   Describe a convenient method for monitoring the reaction rate.

[Answer: the evolution of the nitrogen gas]

   c.   Explain how you would determine the rate law for the reaction.

[Answer: mL $N_2(g)$/sec]

   d.   Write a *generic* rate law for the reaction.   [Answer: rate = $k[NH_4^+]^p[NO_2^-]^q$

4.   Given the following rate data for the reaction, $2 A + 3 C \rightarrow D + 2 E$:

Trial No.	$[A]_o$	$[B]_o$	Rate of Formation of D
1	0.20	0.10	5.0×10^{-6} mol/(L•s)
2	0.30	0.10	7.5×10^{-6} mol/(L•s)
3	0.30	0.30	6.8×10^{-5} mol/(L•s)

   a.   What is the order of the reaction with respect to A?   [Answer: 1st order]

   b.   What is the order of the reaction with respect to B?   [Answer: 2nd order]

   c.   Write the rate law for the reaction.   [Answer: rate = $k [A][B]^2$]

   d.   Calculate the value for the reaction rate constant.

[Answer: $k = 2.5 \times 10^{-3}$ L^2/(mol^2•s)]

5.   How does the rate of a chemical reaction generally change as a function of time?

[Answer: The rate decreases with time because of a decrease in reactant concentration and a buildup of product, increasing the probability of a back-reaction.]

LeChâtelier's Principle; Buffers

INTRODUCTION	Changes in concentration and temperature affect equilibrium positions for a number of chemical systems in this experiment. Other factors affecting the position of chemical equilibria are pH, solubility, gas formation, complex formation, and the common-ion effect. The buffering action of a chemical system is observed when changes in pH are compared when strong acid and strong base are added to water and to a buffer system. Because of the many principles studied, this should be a "core" experiment for any laboratory program.
WORK ARRANGEMENT	Partners. Shared observations and discussions will supplement the experience of shifting equilibria.
TIME REQUIREMENT	3 hours. Do *not* plan to give a quiz during this period. If time is a factor for your laboratory, consider assigning either Parts D and E *or* F and G.

LECTURE OUTLINE

1. Follow the Instruction Routine outlined in "To the Laboratory Instructor."

2. A general discussion of LeChâtelier's principle is appropriate. More specifically, the effects that strong acids and bases have on buffered and unbuffered systems needs to be clearly understood as the student is performing the experiment.

3. Be prepared to discuss each equilibrium that is assigned in the experiment. The multiple equilibria of the Ag^+ ion and the temperature effects in Parts E and G require additional explanations.

4. Each superscript in the Experimental Procedure requires the recording of an observation or equation on the Report Sheet.

5. To avoid crowds at the reagent table, divide the students into three groups at the beginning of the period. Each group should proceed accordingly:
   Group I: Parts A, B, C, D, E, F, G
   Group II: Parts C, D, E, F, G, A, B
   Group III: Parts D, E, F, G, A, B, C

6. A hot water bath is needed for Parts E and G; students can begin preparing this at the beginning of the laboratory period.

CAUTIONS & DISPOSAL

- $6\ M\ HNO_3$, conc NH_3, and conc HCl are used in this experiment. Be prepared to suggest treatment and clean up procedures should these solutions be spilled on the skin or bench top. Have a supply of sodium bicarbonate available for the cleanup of acid and base spills.

- Remind students to properly dispose of the test solutions.

- Discard the test solutions from Parts A, D, E, F, and G in the "Waste Salt Solutions" container.

- Dispose of the silver test solutions, Part B, in the "Waste Silver Salts" container.

TEACHING HINTS

1. Be an inquisitive laboratory instructor; this helps and encourages students to think about equilibria in terms of LeChâtelier's principle.

2. **Part A.** Addition of H_3O^+ shifts the equilibria represented by Equations 24.2 and 24.3 to the *left* because of the reaction of $NH_3(aq) + H_3O^+(aq) \rightarrow NH_4^+(aq) + H_2O(l)$. The solutions to return to the color of the hydrated metal ion.

3. **Part B.** This multiple equilibria system requires students to not only "do" but "think" as each equilibrium is studied; Equations 24.6 through 24.13 should be followed as the tests are made.

B.1. White Ag_2CO_3 precipitates; HNO_3 causes CO_2 evolution.

B.2. HCl precipitates white AgCl which dissolves in conc NH_3, forming colorless $[Ag(NH_3)_2]^+$. Reacidification reprecipitates AgCl.

B.3. KI causes the precipitation of pale yellow AgI.

B.4. Na_2S precipitates black Ag_2S.

4. **Part C.** The buffering action of the $CH_3COOH/CH_3CO_2^-$ system versus the nonbuffering action of water is shown. Refer students to the universal indicator photo on the color plate to estimate of the pH of the solutions. The buffer solution in wells A1 and A2 shows only small pH changes with the addition of 0.1 M NaOH or 0.1 M HCl, whereas with the same addition to water, wells B1 and B2, large changes in pH are observed.

5. **Parts D and E.** $[Co(H_2O)_6]^{2+}$ is pink but turns blue with conc HCl addition due to the formation of $[CoCl_4]^{2-}$. A higher temperature also favors the blue $[CoCl_4]^{2-}$ complex.

6. **Parts F and G.** A concentrated solution of $CuBr_2$ is a deep green-black solution but gradually turns a sky blue with aqueous dilution forming $[Cu(H_2O)_4]^{2+}$. Addition of solid KBr to a dilute copper solution forms the green-black $[CuBr_4]^{2-}$ complex. A high temperature favors the $[Cu(H_2O)_4]^{2+}$ complex.

CHEMICALS REQUIRED	**Part A**		**Part C**	
	0.1 M $CuSO_4$	1 mL	0.1 M CH_3COOH	1 mL
	0.1 M $NiCl_2$	1 mL	0.1 M $NaCH_3CO_2$	1 mL
	conc NH_3 (dropper bottle)	0.5 mL	universal indicator (dropper bottle)	
	1 M HCl (dropper bottle)	1 mL	0.10 M HCl	0.5 mL
			0.10 M NaOH	0.5 mL
	Part B			
	0.01 M $AgNO_3$	0.5 mL	**Parts D and E**	
	0.1 M Na_2CO_3	0.5 mL	1 M $CoCl_2$	1.5 mL
	6 M HNO_3 (dropper bottle)	0.5 mL	conc HCl (dropper bottle)	0.5 mL
	0.1 M HCl	1 mL		
	conc NH_3 (dropper bottle)	0.5 mL	**Parts F and G**	
	0.1 M KI (dropper bottle)	0.5 mL	$CuBr_2(s)$	0.5 g
	0.1 M Na_2S (dropper bottle)	0.5 mL	KBr(s)	0.5 g

SPECIAL EQUIPMENT	24-well plate and Beral pipets	Bunsen burner
	universal indicator chart	"Waste Silver Salts" container
	ring stand and iron rings	"Waste Salt Solutions" container

PRELABORATORY ASSIGNMENT

1. a. If a stress is applied to a system in a state of dynamic equilibrium, the position of the equilibrium shifts to compensate for the applied stress.

   b. A dynamic equilibrium in a system produces no observable changes in the system although the forward and back reactions continue, but at the same rate.

2. a. right   e. left
   b. left   f. left
   c. right   g. right
   d. left

3. a. The dynamic equilibrium exists between the sugar(s) $\rightleftharpoons$ sugar(aq).

   b. A temperature change alters the equilibrium position; in this case the dissolving process is endothermic and a higher temperature shifts the equilibrium right.

4. i ☐d nc

   ☐i d nc

i $\boxed{d}$ nc

$\boxed{i}$ d nc

i $\boxed{d}$ nc

$\boxed{i}$ d nc

i $\boxed{d}$ nc

i $\boxed{d}$ nc

$\boxed{i}$ d nc

$\boxed{i}$ d nc

LABORATORY QUESTIONS

1. a. light blue
   b. deep blue
   c. light green
   d. light blue
   e. pink
   f. white

   g. white
   h. yellow
   i. black
   j. green-black
   k. blue
   l. white

2. a. blue, $[CoCl_4]^{2-}$
   b. blue, $[Cu(H_2O)_4]^{2+}$

3. In the $CH_3COOH/CH_3CO_2^-$ buffer system, concentrations of both H_3O^+ and OH^- are consumed when added; in the HCl/Cl^- system, only concentrations of OH^- are consumed, H_3O^+ is unaffected.

*4. The conjugate base of the buffer system consumes the H_3O^+ of strong acid. When the amount of base has been consumed from the equilibrium system, the buffering action no longer is functional; the *buffer capacity* of the buffer system has been exceeded.

5. $Cr_2O_7^{2-}(aq) + 3 H_2O(aq) \rightleftharpoons 2 CrO_4^{2-}(aq) + 2 H_3O^+(aq)$

LABORATORY QUIZ

1. Explain in Column III (left, right, no change) how the added substance in Column II affects the equilibrium system described in Column I.

Column I	Column II	Column III
$[Cu(H_2O)_4]^{2+}(aq) + 4 NH_3(aq) \rightleftharpoons$ $[Cu(NH_3)_4]^{2+}(aq) + 4 H_2O(l)$	HCl	[left]
$Ag_2CO_3(s) \rightleftharpoons 2 Ag^+(aq) + CO_3^{2-}(aq)$	HNO_3	[right]
$AgCl(s) \rightleftharpoons Ag^+(aq) + Cl^-(aq)$	NH_3	[right]
$[Cu(H_2O)_4]^{2+}(aq) + 4 Br^-(aq) \rightleftharpoons [CuBr_4]^{2-}(aq) + 4 H_2O(l)$	H_2O	[left]
$[Cu(H_2O)_4]^{2+}(aq) + 4 Br^-(aq) \rightleftharpoons [CuBr_4]^{2-}(aq) + 4 H_2O(l)$	heat	[left]
$[Co(H_2O)_6]^{2+}(aq) + 4 Cl^-(aq) \rightleftharpoons [CoCl_4]^{2-}(aq) + 6 H_2O(l)$	HCl	[right]
$[Co(H_2O)_6]^{2+}(aq) + 4 Cl^-(aq) \rightleftharpoons [CoCl_4]^{2-}(aq) + 6 H_2O(l)$	heat	[right]

2. What chemicals are necessary to make up a buffer solution?
   [Answer: a weak acid and its conjugate base or a weak base and its conjugate acid]

3. a. A mixture of acetic acid and sodium acetate forms a buffer. Write an equilibrium equation for this buffer.
   [Answer: $CH_3COOH(aq) + H_2O(l) \rightleftharpoons H_3O^+(aq) + CH_3CO_2^-(aq)$]
   b. What happens to the *mole ratio* of acetic acid to acetate ion as strong base is added to the system? [Answer: the mole ratio decreases]
   c. Explain why sodium acetate alone cannot function as a buffer system.
   [Answer: sodium acetate alone cannot consume added OH^-]

d. What function does the acetic acid/sodium acetate buffer system do that water does not?

[Answer: the buffer resists large changes in pH when a strong acid or strong base is added to the system]

4. At what point in the addition of a strong base does a buffer solution no longer resist changes in its pH?

[Answer: when the buffer capacity is exceeded, i.e., when either the amount of weak acid has been consumed by the strong base]

5. What effect does added heat have on the equilibrium,

$$2 SO_2(g) + O_2(g) \rightleftharpoons 2 SO_3(g) + heat?$$ [Answer: shift left]

6. Identify the color of each:

a. $[Cu(NH_3)_4]^{2+}$ [deep blue]   d. $[Cu(H_2O)_4]^{2+}$ [light blue]
b. AgCl   [white]   e. $[Ni(H_2O)_6]^{2+}$ [light green]
c. Ag_2S [black]   f. $[CoCl_4]^{2-}$ [blue]

Experiment 25
An Equilibrium Constant

INTRODUCTION

The equilibrium constant for the $Fe^{3+}(aq) + SCN^-(aq) \rightleftharpoons FeNCS^{2+}(aq)$ equilibrium is determined spectrophotometrically. For most students this is the first time an instrument other than a balance is used in the laboratory; therefore, detailed instructions about its operation is absolutely necessary.

WORK ARRANGEMENT

Partners. The grouping of students depends on the availability of spectrophotometers and on how quickly students can cycle on and off the spectrophotometer.

TIME REQUIREMENT

3 hours. 2 hours are required to collect the data and 1 hour should be allotted for the discussion and calculations.

LECTURE OUTLINE

1. Follow the Instruction Routine outlined in "To the Laboratory Instructor."

2. Describe the chemical system being studied in the experiment and how the equilibrium constant is to be measured.

3. A general discussion of the workings of a spectrophotometer, transmittance and absorbance of EM radiation, and concentration of absorbing specie help to elevate the understanding of the spectrophotometer from a "black box" to an analytical instrument.

4. Students are to understand the difference between the position of equilibrium in the set of standard solutions (Part A, Table 25.2) versus that in the set of test solutions (Part B, Table 25.3).

5. **Part A.** The position of equilibrium for the standard solutions is shifted toward the formation of $FeNCS^{2+}$ because of the large excess of Fe^{3+} in the system (Figure 25.6). (Nearly) all of the SCN^- is assumed to be in the form of the $FeNCS^{2+}$ complex at equilibrium; the moles of $FeNCS^{2+}$ is assumed to equal the original moles of SCN^-.

6. Because the linear %T scale on the spectrophotometer is easier to interpolate than is the logarithmic absorbance scale, we advise students to record the %T of their solutions and calculate the corresponding absorbance value for the solution.

7. **Part B.** An understanding of the Beer's law plot in Part A is necessary for the analysis of the data in Part B. Remember that the $[FeNCS^{2+}]$ at equilibrium (Part A) equals the moles of SCN^- in 100 mL of solution (see Teaching Hint 6).

8. The analysis of the data is extensive and detailed. Students may need some assistance with the calculations.

9. Class data for the value of K_c are to be collected at the end of the experiment. Advise students as to the procedure by which they are to collect this data from other student groups.

CAUTION & DISPOSAL

• The chemicals used in this experiment are safe to handle and produce no environmental dangers. We provide a "Waste Salts" container for the disposal of the test solutions in Parts A and B.

TEACHING HINTS

1. Give students detailed instructions on the operation of the spectrophotometer. We in addition place the following (next page) set of printed instructions beside the spectrophotometer:

Instructions for Operating a Spectrophotometer

1. Turn on the instrument and allow a 10 minute warm up period. (You can do this before presenting your lecture.)

2. Turn the wavelength control of the monochromator to the desired wavelength (447 nm or at a maximum absorption).

3. Turn the zero control knob to read 0%T with an empty sample compartment and the sample compartment lid closed.

4. Place the "blank" solution into a cuvet (Part A, Solution 1), insert the cuvet into the sample compartment, close the lid to the sample compartment, and adjust to 100%T with the reference (or 100%T) control knob.

5. Place a sample solution into a cuvet, insert it into the sample compartment, close the lid, and record the %T, the percent of the light that is transmitted through the solution.

2. **Part A.1.** To conserve 100-mL volumetric flasks, several student groups can use the same set of standard solutions. Each group should take its own %T readings and establish its own calibration curve (Beer's law plot).

3. **Part A.1.** Advise students to use rubber bulbs for pipetting solutions. Caution students not to mix pipets between solutions.

4. **Part A.2, 3.** Instruct students on the proper handling of a cuvet; handling only the lip of the cuvet and drying the outside with a lint-free towel, e.g., Kimwipes. The orientation of the cuvet should be the same each time the cuvet is placed into the sample compartment.

5. **Part A.4.** The cuvets must be clean, rinsed several times with the test solution, and dried on the outside before each measurement.

6. **Part A.5.** Some assistance may be required for the calculation of the [FeNCS^{2+}] at equilibrium—*remember* mol SCN$^-$ initially equals mol FeNCS^{2+} at equilibrium (see Figure 25.6). Students must draw the *best straight* line through the data points and the origin.

7. **Part B.** Make sure students record the *exact* molar concentrations of the solutions.

8. **Part B.1, 2.** The [Fe^{3+}] and [SCN$^-$] are nearly the same in the test solutions (Figure 25.7). Be sure students use the correct (i.e., 0.002 M) Fe(NO$_3$)$_3$ solution in Part B. Students must stir the solution to reach an equilibrium. Repeated re-calibration of the spectrophotometer is advisable for the reproducibility of the data.

9. **Part B.3.** After completing the experiment, the volumetric flasks, pipets, and cuvets must be rinsed with tap and deionized water and allowed to air-dry.

10. **Part B.4.** The Beer's law plot from Part A must be correctly interpreted for the analysis of the data in Part B. Students often do not tie in the relationship of Parts A and B.

11. **Part B.5.** For assistance in the calculations I offer a representative set of data:

Part A (0.00269 M NaSCN was used for this data)

Solution	1	2	3	4	5
[FeNCS^{2+}]	0	5.38×10^{-5}	1.08×10^{-4}	1.61×10^{-4}	2.15×10^{-4}
Absorbance	0	0.242	0.50	0.75	0.99

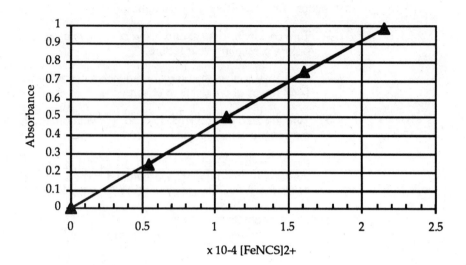

x 10-4 [FeNCS]2+

Part B (0.00293 M Fe(NO$_3$)$_3$ and 0.00269 M NaSCN were used for this data)

Solution	1	2	3	4	5
Absorbance	0.33	0.46	0.70	0.87	1.09

Calculations

[FeNCS^{2+}], from calibration curve

7.11×10^{-5}	1.00×10^{-4}	1.52×10^{-4}	1.88×10^{-4}	2.33×10^{-4}

moles FeNCS^{2+}, at equilibrium (volume = 10 mL)

7.11×10^{-7}	1.00×10^{-6}	1.52×10^{-6}	1.88×10^{-6}	2.33×10^{-6}

[Fe^{3+}], equilibrium

moles Fe^{3+}, reacted (same as moles of FeNCS^{2+} formed)

7.11×10^{-7}	1.00×10^{-6}	1.52×10^{-6}	1.88×10^{-6}	2.33×10^{-6}

moles Fe^{3+}, unreacted (initial moles – reacted moles); initial moles = 1.47×10^{-5} moles Fe^{3+}

1.40×10^{-5}	8.95×10^{-6}	8.44×10^{-6}	8.07×10^{-6}	7.62×10^{-6}

[Fe^{3+}], equilibrium, unreacted (÷ by 10 mL or 0.010 L)

1.40×10^{-3}	8.95×10^{-4}	8.44×10^{-4}	8.07×10^{-4}	7.62×10^{-4}

[SCN$^-$], equilibrium

moles SCN$^-$, reacted (same as moles of FeNCS^{2+} formed)

7.11×10^{-7}	1.00×10^{-6}	1.52×10^{-6}	1.88×10^{-6}	2.33×10^{-6}

moles SCN$^-$ unreacted (initial moles – reacted moles)

1.98×10^{-6}	4.38×10^{-6}	6.56×10^{-6}	8.88×10^{-6}	1.11×10^{-5}

[SCN$^-$], equilibrium, unreacted (÷ by 10 mL or 0.010 L)

	1.98×10^{-4}	4.38×10^{-4}	6.56×10^{-4}	8.88×10^{-4}	1.11×10^{-3}
K_c	256	255	274	262	275

Ave K_c	264
Average deviation:	7.6
Standard deviation:	9.2

CHEMICALS REQUIRED

It is suggested that four student groups use one spectrophotometer and one set of standard solutions. However, the equilibrium solutions for Part B should be prepared by *each* group.

0.2 M Fe$(NO_3)_3$ in 0.25 M HNO$_3$*	125 mL for 4 groups
0.002 M NaSCN in 0.25 M HNO$_3$* (prepare daily)	65 mL for 4 groups
0.25 M HNO$_3$	400 mL for 4 groups
0.002 M Fe$(NO_3)_3$ in 0.25 M HNO$_3$*	25 mL for *each* group

*The molar concentrations of these solutions should be known to three significant figures; the stock reagent bottles should be labeled accordingly.

SPECIAL EQUIPMENT

Spectrophotometer (Check it out—make sure it works properly before the laboratory session begins!)

spectrophotometer	1 per 4 groups
cuvets	5 per 4 groups
lint-free towel (Kimwipes)	
100-mL volumetric flasks	5 per 4 groups in Part A
10-mL (1 mL graduations) pipet	1 per 4 groups in Part A
25-mL pipet and pipet bulb	1 per 4 groups in Part A
150-mm test tubes *or* 10-mL volumetric flasks	5 per group in Part B
1-mL pipet (optional)	1 per group for Part B
5-mL pipet	1 per group in Part B
5-mL (0.1-mL graduations) pipet	2 per group In Part B
extra rubber pipet bulb	1 per group
"Waste Salts" container	

PRELABORATORY ASSIGNMENT

1. The energy that is absorbed by the molecule excites an electron from a lower to a higher energy state.

2. • the thickness of the solution through which the light passes
   • the concentration of the absorbing species in solution
   • the absorptivity coefficient of the absorbing species

3. a. Near 600 nm (Table 25.1)
   b. Blue EM radiation is absorbed at 447 nm and the orange (orange-yellow) EM radiation is transmitted.

4. A large excess of Fe^{3+} causes the equilibrium in Equation 25.5 to shift far to the right, essentially consuming all of the SCN$^-$ in forming the FeNCS^{2+} complex ion (see Figure 25.6).

5. A = 0.66

6. a. 1.0×10^{-4} mol Fe^{3+} f. 9.3×10^{-5} mol Fe^{3+} remaining unreacted
   b. 1.0×10^{-5} mol SCN$^-$ g. 3.0×10^{-6} mol SCN$^-$ remaining unreacted
   c. 7.0×10^{-6} mol FeNCS^{2+} h. 9.3×10^{-3} mol/L = [Fe^{3+}] at equilibrium
   d. 7.0×10^{-6} mol Fe^{3+} reacted i. 3.0×10^{-4} mol/L = [SCN$^-$] at equilibrium
   e. 7.0×10^{-6} mol SCN$^-$ reacted   j. 7.0×10^{-4} mol/L = [FeNCS^{2+}] at equilibrium
                                          k. $2.5 \times 10^2 = K_c$

7. $[SCN^-] = \dfrac{[FeNCS^{2+}]}{K_c\,[Fe^{3+}]} = \dfrac{1.0 \times 10^{-6}}{2.5 \times 10^2\,[0.090]} = 4.4 \times 10^{-8}$ mol/L

8. Because the $Fe^{3+}(aq)$ in 0.25 M HNO_3 may have a very slight absorbance at 447 nm; its contribution to absorbance is canceled when it is used as a blank to "zero" the instrument (Part A, Solution 1).

LABORATORY QUESTIONS

1. a. Less light is transmitted to the detector, causing a smaller %T.
   b. *Too high.* A smaller %T corresponds to a larger absorbance and therefore a higher $[FeNCS^{2+}]$. From Equation 25.6, the larger the $[FeNCS^{2+}]$ the larger K_c.

2. The thickness of the solution and the molar absorptivity of $FeNCS^{2+}$ are not considered because the thickness of the cuvet is assumed constant (especially if the same cuvet is used for all measurements) and the probability of photon absorption is assumed constant (especially if the same wavelength is used for all measurements).

3. A low %T implies that the concentration of the absorbing specie is very high; to increase the %T the sample can be further diluted with 0.25 M HNO_3.

4. a. absorbance readings increase   d. lower $[SCN^-]$
   b. higher $[FeNCS^{2+}]$ e. higher K_c
   c. lower $[Fe^{3+}]$

LABORATORY QUIZ

1. Copper ion solutions appear blue to the eye. Is this the absorbed or transmitted light? Explain.   [Answer: transmitted light]

2. Characterize an electronically "excited state" of an atom or molecule.
   [Answer: an electron has absorbed energy to acquire a higher energy state—the atom is in an excited state]

3. Identify two factors that control the amount of electromagnetic energy that a sample absorbs.   [Answer: the thickness of the sample and the absorptivity coefficient]

4. The meter on the spectrophotometer reads both the absorbance and the percent transmittance.
   a. Which scale is linear?   [Answer: %T]
   b. Which parameter is monitored by the photocell?   [Answer: transmitted light]

5. Beer's Law is A = a•b•c. Identify each term in the Beer's law equation.
   [Answer: see answer to Prelab Question 2 or see pages 305–6 in the manual]

6. In the equilibrium, $Fe^{3+}(aq)$ + $SCN^-(aq)$ ⇌ $FeNCS^{2+}(aq)$, a large excess of Fe^{3+} is added to an accurately measured sample of SCN^-. Is the $[FeNCS^{2+}]$ at equilibrium approximately the same as the initial $[Fe^{3+}]$ or the initial $[SCN^-]$? Explain.
   [Answer: SCN^- concentration]

7. a. Given the following experimental equilibrium constants, calculate the average deviation and the standard deviation for a reported equilibrium constant derived from this data:

Trial No.	K_c
1	14.0
2	17.0
3	18.0
4	15.0
5	19.0

   [Answer: average deviation = 1.7; standard deviation = 2.1]
   b. Sketch error curve that represents the standard deviation range for this set of data.

INTRODUCTION

A titrimetric analysis is used to determine the "strength" of an antacid. The principles associated with acids and bases and terms such as pH, buffer, and molar concentration are a part of the analysis. You, as an instructor, need to be well-prepared and anticipate the questions associated with the procedure . . . but, it's fun.

A standard 0.1 M NaOH solution is required for the analysis of the antacid. Either the student must prepare the NaOH solution as described in Experiment 9 or the stockroom must prepare the solution in advance of the experiment.

WORK ARRANGEMENT

Individuals. The quantitative nature of the experiment is a valuable learning experience for "new" chemists.

TIME REQUIREMENT

3 hours for the four antacid titrations (2 trials for each of 2 antacids). The time requirement is also dependent upon the availability of the standard 0.1 M NaOH solution. To reduce the laboratory time, consider the analysis of only one antacid but require three trials with a good precision of data.

LECTURE OUTLINE

1. Follow the Instruction Routine outlined in "To the Laboratory Instructor."

2. Discuss the applicability of chemistry to everyday living and more specifically to antacids.

3. The techniques for the quantitative transfer of a liquid sample (especially Technique 16) should be reviewed. Chemists "think and do" quantitatively; since it has been awhile since the last quantitative titrimetric analysis, students need to review the prescribed techniques in Technique 16C. Review the procedure for cleaning a buret and delivering titrant.

4. Review the correct technique for pipetting a liquid (Technique 16B).

5. Excess, standardized HCl is used to destroy any buffer action in the antacid and to neutralize the base of the antacid; the solution is boiled to remove the CO_2 (CO_2 is an acidic anhydride) that is generated in the neutralization (but only if the base is a carbonate or bicarbonate in the antacid); the unreacted HCl is titrated to the bromophenol blue endpoint with the standardized NaOH. This back-titration procedure, the titration of an excess of a known reagent to the chemical system, is new a new concept for students. Students may require assistance in completing the calculations.

6. Students are to complete 2 trials for each antacid; encourage students to analyze their own antacid.

7. Demonstrate the appearance of the bromophenol blue endpoint to the class. Bromophenol blue is blue at a pH < 3.0 and blue at pH > 4.6.

CAUTION & DISPOSAL

- The solutions used in this experiment can be safely discarded in the sink, followed by a generous supply of tap water. Check with your local safety officer to be certain that local disposal policies are being followed.

TEACHING HINTS

1. **Part A.2.** Heat removes CO_2 from the reaction mixture. Carbon dioxide is acidic and, therefore, its presence will affect the amount of NaOH titrant that would be dispensed in reaching the bromophenol blue endpoint.

2. **Part A.2.** If the solution is blue when bromophenol blue is added, another 10-mL aliquot of 0.1 M HCl must be added (the student must be analyzing a strong antacid if more HCl is needed!).

3. **Part A.2.** The excess HCl is used to exceed any buffering capacity in the antacid. In addition, a low pH endpoint is used to avoid any re-establishment of a buffer solution. The pH range of the color range for bromophenol blue is 3.0 (yellow) to 4.6 (blue).

4. **Part B.1.** The buret is to be properly prepared and the volume of the 0.1 M NaOH titrant is properly recorded.

5. **Part B.2.** *Again* emphasize the importance of good titration technique. It is worth your time and the students' precision of data to use proper titration techniques.

6. **Data Information:** A 1.8-g Maalox tablet has 350 mg $Mg(OH)_2$ and 350 mg $Al(OH)_3$. The 350 mg of $Mg(OH)_2$ requires 120 mL of 0.1 M HCl and the 350 mg $Al(OH)_3$ requires 135 mL of 0.1 M HCl for neutralization. Therefore a 0.2-g sample of Maalox requires about 29 mL of 0.1 M HCl.

$$0.350 \text{ g Mg(OH)}_2 \times \frac{\text{mol Mg(OH)}_2}{58.3 \text{ g Mg(OH)}_2} \times \frac{2 \text{ mol H}^+}{1 \text{ mol Mg(OH)}_2} \times \frac{L}{0.1 \text{ mol HCl}}$$
$$= 0.120 \text{ L of } 0.1 \text{ } M \text{ HCl}$$

A 1.5-g Rolaids tablet has 300 mg of $NaAl(OH)_2CO_3$, requiring 83.3 mL of 0.1 M HCl for neutralization. Therefore, a 0.2-g sample requires about 12 mL of 0.1 M HCl.

CHEMICALS REQUIRED	0.1 M HCl (standardized) 150 mL 0.1 M NaOH (standardized) prepared according to the procedure in Experiment 9 75 mL bromophenol blue indicator (dropper bottle) 2 antacids (commercial) 2 g
SUGGESTED UNKNOWNS	Commercial antacids can be used. Laboratory antacids can be prepared by mixing $CaCO_3$, $NaHCO_3$, or MgO with an inert chemical, such as NaCl. The student uses a mass of sample that requires about 5–15 mL of standardized 0.1 M NaOH for neutralizing the excess HCl..
SPECIAL EQUIPMENT	Erlenmeyer flasks, 250-mL 2 | 25-mL pipet and pipet bulb 1 balance (±0.001 g) | 10-mL pipet 1 mortar and pestle | ring stand and buret clamp 1 50-mL buret and buret brush 1
PRELABORATORY ASSIGNMENT	1. $NaAl(OH)_2CO_3(aq) + 4 H^+(aq) \rightarrow Na^+(aq) + Al^{3+}(aq) + 3 H_2O(l) + CO_2(g)$

2. The pH range for color change for bromophenol blue is 3.0 to 4.6.
   a. Color in a solution with a pH < 3: yellow
   b. Color in a solution with a pH > 4.6: blue
   c. yellow to blue

3. The antacids containing carbonates or bicarbonates will, in the presence of excess H_3O^+, produce CO_2 gas (Equation 26.4). This causes CO_2 gas to accumulate on the stomach and causes one to "belch."

4. 1.70×10^{-3} mol base in the antacid.

5. 0.095 g $Al(OH)_3$ requires 3.65×10^{-3} mol H_3O^+ and 0.412 g $MgCO_3$ requires 9.77×10^{-3} mol H_3O^+ for neutralization or a total of 1.34×10^{-2} mol H_3O^+.

6. 0.450 g $Mg(OH)_2$ requires 1.54×10^{-2} mol H_3O^+ and 0.500 g $Al(OH)_3$ requires 1.92×10^{-2} mol H_3O^+ for neutralization or a total of 3.47×10^{-2} mol H_3O^+.

LABORATORY QUESTIONS	1. *More* NaOH dispensed. CO_2 in water is acidic; NaOH is required to neutralize the excess HCl *and* the CO_2 before the bromophenol blue endpoint is reached.

2. Milk of magnesia contains no carbonate or bicarbonate. When excess HCl is added, no CO_2 forms; therefore the heating step in Part A.2 may be omitted.

3. a. $C_6H_5O_7^{3-}(aq) + 3\,H_3O^+(aq) \rightarrow H_3C_6H_5O_7(aq) + 3\,H_2O(l)$
   b. $Mg(OH)_2(s) + 2\,H_3O^+(aq) \rightarrow Mg^{2+}(aq) + 2\,H_2O(l)$
   One mole of sodium citrate neutralizes three moles of H^+, whereas one mole of magnesium hydroxide neutralizes only two moles of H^+.
   Therefore sodium citrate is more effective *per mole*.

   For sodium citrate (molar mass = 258g/mol), the effectiveness is 1 mol H^+/86.0 g $Na_3C_6H_5O_7$
   For magnesium hydroxide (molar mass = 58.3 g/mol), the effectiveness is 1 mol H^+/29.2 g $Mg(OH)_2$
   Therefore, $Mg(OH)_2$ is more effective *per gram*.

4. $0.500 \text{ g CaCO}_3 \times \dfrac{\text{mol CaCO}_3}{100.1 \text{ g CaCO}_3} \times \dfrac{2 \text{ mol HCl}}{1 \text{ mol CaCO}_3} \times \dfrac{36.5 \text{ g HCl}}{\text{mol HCl}} = 0.365 \text{ g HCl}$

5. $1.0 \text{ L} \times \dfrac{(1.0 \times 10^{-1} - 1.0 \times 10^{-2}) \text{ mol H}^+}{\text{L}} \times \dfrac{1 \text{ mol CaCO}_3}{2 \text{ mol HCl}} \times \dfrac{100.1 \text{ g CaCO}_3}{\text{mol CaCO}_3}$
   $= 4.50 \text{ g CaCO}_3$

LABORATORY QUIZ 1. Since antacids are bases, why aren't antacids titrated directly with a standardized acid solution rather that adding an excess of standardized acid solution, followed by a back titration with standardized base solution?
   [Answer: because of the buffering action present in many antacids; the excess standardized acid removes the buffer action]

2. A 50-mL volume of 0.107 M HCl is added to an antacid sample. The solution is titrated to the bromophenol blue endpoint using 23.7 mL of 0.0971 M NaOH. How many moles of base are in the antacid sample?   [Answer: 3.05×10^{-3} mol or 3.05 mmol base]

3. If the bromophenol blue endpoint is surpassed in the titration of the antacid with the standardized NaOH solution, are the moles of base in the antacid reported too high or too low? Explain   [Answer: low]

Vinegar Analysis

INTRODUCTION	A standardized NaOH, prepared by stockroom personnel or according to the procedure of Experiment 9, is used for the vinegar analysis. Because of the "relevance" of this experiment, students enjoy collecting data and comparing results. To heighten interest we encourage students to bring their own vinegar sample.
WORK ARRANGEMENT	Individuals.
TIME REQUIREMENT	2 hours for 2 trials on two vinegars (4 trials total) including calculations, *if* the standardized NaOH solution is already available.

LECTURE OUTLINE

1. Follow the Instruction Routine outlined in "To the Laboratory Instructor."

2. Briefly review the theory and experimental procedure as the principles of an acid-base titration should, by now, be very straight forward.

3. Students should review Technique 16 and Experiment 9, Parts A.4 and A.5 before beginning the experiment.

4. A calculation in Part 1 is required before the experiment is to be performed. The calculated volume of 5% vinegar to used for the reaction of 25 mL of 0.1 M NaOH is about 3 mL.

   $$0.025 \text{ L} \times \frac{0.1 \text{ mol NaOH}}{\text{L}} \times \frac{1 \text{ mol CH}_3\text{COOH}}{1 \text{ mol NaOH}} \times \frac{60.0 \text{ g}}{\text{mol}} \times \frac{100 \text{ g soln}}{5 \text{ g CH}_3\text{COOH}}$$

   = 3 g solution or 3 mL of solution, assuming a density of 1 g/mL

5. All traces of the NaOH titrant must be removed from the buret after the analysis. A thorough cleanup of all glassware is necessary.

CAUTIONS & DISPOSAL

- If students are to standardize their own NaOH solution, according to the Experimental Procedure in Experiment 9, then the concentrated NaOH solution must be handled carefully.

- All test solutions can be discarded in the sink, followed by a generous flow of tap water.

TEACHING HINTS

1. A wash bottle filled with boiled, deionized water should be available.

2. For some dark vinegars the phenolphthalein endpoint is difficult to detect; encourage students to titrate white vinegars.

3. Refer students to the color plate for the appearance of phenolphthalein at the endpoint.

4. Require students to use their "best" titration techniques; you should grade accordingly.

CHEMICALS REQUIRED	0.1 M NaOH (standardized) prepared according to the procedure in Experiment 9 or by stockroom personnel 　　　　　　　　　　　　　　　　125 mL Phenolphthalein indicator 1 dropper bottle/10 students
SUGGESTED UNKNOWNS	Vinegar, commercial, or prepare several 3 to 5% acetic acid solutions for unknowns　　10 mL

SPECIAL EQUIPMENT				
125- or 250-mL Erlenmeyer flasks	2	balance (±0.01 g)		
50-mL buret and buret brush	1	ring stand and buret clamp		1

PRELABORATORY ASSIGNMENT

1. a. 2.86×10^{-3} mol CH_3COOH
   b. 0.172 g CH_3COOH
   c. 4.19% CH_3COOH

2. Phenolphthalein changes from colorless to pink in the analysis of the vinegar.

3. No water droplets cling to a clean glass surface.

4. The wait for 30 s before reading a buret allows time for the titrant to drain from the wall of the buret.

*5. Since acetic acid is a weak acid, the pH at its stoichiometric point in a titration with NaOH is greater than 7, in the same pH range as the color change for phenolphthalein. Methyl orange changes color will below a pH of 7 and therefore a "false" stoichiometric point would be recorded in the titration. For a strong acid-NaOH titration, the break in the pH curve occurs sharply over the pH range of about 4 to 9 and therefore either phenolphthalein or methyl orange would be suitable indicators.

LABORATORY QUESTIONS

1. *Too high.* The misplaced drop is recorded as standard NaOH solution dispensed from the buret, indicating a greater number of moles of acetic acid in the vinegar. This calculates to a greater percent acetic acid.

2. a. A mass measurement of ±0.01 g on a balance is more accurate than a volume of only ±0.02 mL. In addition, the experiment requests that the composition of the vinegar be reported as percent *by mass*.
   b. The density of the vinegar would have to be determined if the amount is measured volumetrically.

3. a. The more phenolphthalein that is used in the analysis (it too, is an acid) the more the standard NaOH solution that is required for the neutralization of the acetic acid.
   b. *Too high.* With more moles of NaOH dispensed from the buret, the moles and the percent acetic acid would calculate to be too high.

LABORATORY QUIZ

1. How does the amount of phenolphthalein in the vinegar affect the quantity of standardized NaOH needed in the analysis of a vinegar sample?
   [Answer: since phenolphthalein is also a weak acid, additional NaOH would be needed]

2. A 17.7-mL volume of 0.111 *M* NaOH solution is required to reach the phenolphthalein endpoint in titrating 2.63 g of vinegar. Calculate the percent by mass of acetic acid in the vinegar. [Answer: 4.49%]

3. If the endpoint is surpassed with the standard NaOH titrant, is the reported percent by mass of acetic acid in vinegar high or low? Explain. [Answer: high]

Potentiometric Analyses

INTRODUCTION   Students use a pH meter in the analyses described in this experiment; therefore careful instruction and supervision is *absolutely* necessary. Many principles other than pH are covered in this experiment: students should have titration experience, a "hands-on" knowledge of pH, graphing experience, and be well-informed on acid-base theory *before* attempting this experiment.

A standardized NaOH solution is required for this experiment—either the student must prepare the solution according to the procedure described in Experiment 9, or the solution must be prepared in the stockroom.

WORK ARRANGEMENT   Partners, depending on the availability of pH meters.

TIME REQUIREMENT   3 hours (assuming the standardized NaOH solution is already prepared): 2 hours for the 6 titrations, 1 hour for graphs and calculations.

LECTURE OUTLINE

1. Follow the Instruction Routine outlined in "To the Laboratory Instructor."

2. Discuss the theory of a pH (or $[H_3O^+]$) measurement with a pH meter. Provide instructions for operating a pH meter; a demonstration of its operation is suggested.

3. Review the three measurements (molar concentration of a weak acid and the molar mass and pK_a of a solid weak acid) to be completed in the experiment and also the corresponding graphical data that are required for each.

4. Discuss the construction of the titration curve in Figure 28.3 and the determination of the pK_a of a weak acid from a titration curve (Figure 28.4).

5. Describe how students can better budget their time for this experiment. Because of its length of the experiment, you may want to reduce the number of trials for Parts B and C to 2 trials. On occasion, we allow students to take their data home to plot the data for the analyses.

6. Make assignments of pH meters to student groups.

7. Students should review Technique 16C and Experiment 9, Part A.4 and A.5 *before* beginning the experiment.

8. At the conclusion of the experiment, the buret and glassware should be rinsed clean with several portions of tap water and a final rinse with deionzed water.

9. Class data are to be collected for the pK_a of each weak acid at the conclusion of the data analysis. A consensus pK_a and a standard deviation for its measurement is to be calculated. Describe to students how this is to be done.

CAUTION & DISPOSAL   •   All test solutions can be discarded in the sink, followed by a flush with tap water.

TEACHING HINTS

1. **Part A.** You can save laboratory time by turning on the pH meters and checking their operation before your lab lecture.

    Insist on the proper care of the pH meter and the measuring electrodes—require students to repeat Part A.2 every time the electrodes are removed from the solution. When the electrodes are removed from the solution, return the pH meter to "Standby", rinse the electrodes with deionized water and, if not to be used immediately, submerge the electrodes in deionized water.

2. **Part B.1.** To save laboratory time, the standard 0.1 *M* NaOH can be prepared in advance in the stockroom.

3. **Part B.4.** With the addition of each increment of NaOH titrant, the solution in the beaker is to be swirled, and time must be allowed for the pH reading on the pH meter to stabilize. This is especially true as the reaction mixture nears the stoichiometric point. The pH versus volume of NaOH data is to be recorded on a separate sheet of paper. Deionized water from a wash bottle should *not* be used to wash the sides of the beaker; the deionized water dilutes the solution and therefore affects the pH measurement of the solution.

4. **Part B.6.** Approve the plotted titration curve. Refer to Appendix C for the proper construction and labeling of a graph.

5. **Part B.7.** Some assistance with the calculations may be necessary.

6. **Part C.1.** Ethanol may be added to the solution if the solid acid sample is slow to dissolve.

7. **Part C.** Students must be informed of the type of acid (monoprotic or diprotic) they have been issued as an unknown. This analysis follows the same procedure as in Part B. The pK_a is determined from the pH versus V_{NaOH} graph (see Figure 28.4).

CHEMICALS REQUIRED	standard pH buffer solutions (pH = 4.0 and/or 7.0) 100 mL/pH meter 0.1 *M* NaOH (standardized) prepared according to the procedure in Experiment 9 or by stockroom personnel 200 mL/student group 95% ethanol (optional) 25 mL
SUGGESTED UNKNOWNS	**Part B.** Various standard acetic acid solutions varying from 0.05 *M* to 0.150 *M* (this concentration range requires from 12.5 mL to 37.5 mL of 0.1 *M* NaOH for neutralization) ≈ 90 mL

Part C. The solid acid samples should be dried in a drying oven for about 1 hour at 110°C prior to a mass measurement.

Acid	Formula	Acid Type	Molar Mass	pK_a*
adipic acid	HOOC-(CH2)4-COOH	H2A	146.14	4.43; 5.41
citric acid	HOOC-C(OH)-(CH2-COOH)2	H3A	192.1	3.14; 4.77; 6.39
d,l-mandelic acid	C6H5CH(OH)-COOH	HA	152.2	3.85
fumaric acid (trans)	HOOC-CH=CH-COOH	H2A	116.1	3.03; 4.44
maleic acid (cis)	HOOC-CH=CH-COOH	H2A	116.1	1.83; 6.07
malonic acid	HOOC-CH2-COOH	H2A	104.1	2.83; 5.69
oxalic acid	HOOC-COOH	H2A	90.04	1.23; 4.19
phthalic acid	HOOC-C6H4-COOH	H2A	166.0	2.89; 5.51
potassium hydrogen phthalate	HOOC-C6H4-COO⁻K⁺	HA	204.2	5.51
potassium hydrogen tartrate	HOOC-(CHOH)2-COO⁻K⁺	HA	188.1	4.34
sodium hydrogen oxalate	HOOC-COO⁻Na⁺	HA	112.0	4.19
sodium hydrogen sulfate	NaHSO4	HA	120.1	1.92
succinic acid	HOOC-(CH2)2-COOH	H2A	118.0	4.16; 5.61
sulfamic acid	H2N-SO3H	HA	97.1	1
tartaric acid	HOOC-(CHOH)2-COOH	H2A	150.1	2.98; 4.34

*pK_a values are from *Handbook of Chemistry and Physics*, 66th Edition, Robert C. Weast, Editor, CRC Press, Inc. Boca Raton, Florida, 1985.

SPECIAL EQUIPMENT	pH meter and pH electrodes	1/group	weighing paper	3
	Kimwipes	1 box/pH meter	balance (±0.001 g)	
	50-mL buret and buret brush	1/pH meter	ring stand and buret clamp	1
	25-mL pipet and pipet bulb	1	additional graph paper	
	drying oven			

PRELABORATORY ASSIGNMENT

1.   a.

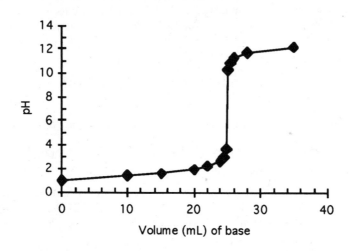

   b.   25.0 mL

2.   a.   A specific ion electrode is sensitive to a difference in the concentration of a particular ion on two sides of a membrane—the differences in concentration is detected as an electrochemical potential.
   b. H_3O^+ ion is the specific ion measured in this experiment. The potential difference between two concentrations of H_3O^+ across a glass membrane is measured against that of the standard calomel electrode (S.C.E.).

3.   0.0924 M HNO_2

4.   a.   The potential, established at the interface of the glass electrode, varies with the $[H_3O^+]$ in solution. This potential is measured relative to the standard calomel electrode. The potential difference between the two electrodes is displayed directly as pH even though it is a potential that the instrument is recording.
   b.   *No*. The equilibrium concentrations of the substances that constitute the calomel electrode are unaffected by any stresses placed on the calomel system; the amounts of Hg_2Cl_2, Hg, and Cl^- remain unchanged for the pH measurements.
   $Hg_2Cl_2(s) + 2e^- \rightleftharpoons Hg(l) + 2Cl^-(aq, satd)$
   c.   *Yes*. The Ag-AgCl redox couple is sensitive to the availability of H_3O^+ in the 0.1 M HCl solution which, in turn, is sensitive to the $[H_3O^+]$ on the outside of the glass electrode. This change in the Ag-AgCl redox potential causes the potential difference (and thus, changes in pH) to be detected when changes in $[H_3O^+]$ occur in solution.

LABORATORY QUESTIONS

1.   *High*. If the stoichiometric point is misread to be at a larger volume, then more moles of acid are recorded as being neutralized causing a higher molar concentration of the acid.

2.   a.   *Unchanged*. The molar concentration of the acid remains unchanged because the volume of base needed to reach the stoichiometric point (the inflection point in the titration curve) is what is used to calculate molar concentration, not the specific pH.
   b.   *Unchanged*. The molar mass of the weak acid is also unaffected. The mass of the acid and the stoichiometric point on the titration curve are used to calculate the molar mass.
   c.   *Higher*. The pK_a of the weak acid, read directly from the titration curve, is

dependent upon the observed pH reading, thus calibration of the pH affects the determination of the pK_a.

LABORATORY QUIZ

1. Construct the *general* shape for the following graphs:
   a. pH versus V_{NaOH} for the titration of a strong acid with a strong base
   b. pH versus V_{NaOH} for the titration of a weak acid with a strong base

2. For the titration of a weak acid with a strong base, describe how the pK_a of the weak acid is determined.   [Answer: pK_a = pH at the halfway stoichiometric point]

3. Describe briefly the preparation of a standardized NaOH solution. [Answer: See Expt 9]

4. A 27.5-mL volume of a 0.107 M NaOH solution is required to reach the stoichiometric point in titrating 25.0 mL of an HCl solution with unknown concentration. What is the molar concentration of the HCl solution?   [Answer: 0.118 M HCl]

5. The pH at the stoichiometric point for the titration of
   a. a strong acid with a strong base is (=7, >7, or <7).   [Answer: = 7]
   b. a weak acid with a strong base is (=7, >7, or <7).   [Answer: > 7]

6. What is the name of the reference electrode used for most pH meters?
                    [Answer: standard calomel electrode]

Experiment 29
Aspirin Synthesis and Analysis

INTRODUCTION

An interesting experiment for students. A familiar compound is synthesized and then analyzed for its percent purity. Good laboratory technique and safety practices are required to obtain a product of high yield and high purity. The only drawback to this experiment is that students may not fully understand the nature of the organic chemicals used in the synthesis.

A standardized NaOH solution is required for Part C of this experiment—either the student must prepare the solution according to the procedure described in Experiment 9 or the solution must be prepared in the stockroom.

WORK ARRANGEMENT

Individuals. The students should practice good laboratory technique and be safety conscious.

TIME REQUIREMENT

3 hours, *if* the standardized 0.1 *M* NaOH solution is already prepared.

LECTURE OUTLINE

1. Follow the Instruction Routine outlined in "To the Laboratory Instructor."

2. Review the chemistry of the preparation of the aspirin. Students may not be know and/or understand the chemistry of the organic compounds, but they should understand the synthesis of the aspirin and the acid-base chemistry of the analysis.

3. Yield and purity of the product are factors to evaluate in determining a grade for this experiment. Both purity checks (the melting point and the titrimetric analysis) should be completed on the product.

4. Extra emphasis on safety, regarding the organic chemicals, should be discussed with the students.

5. A review of the vacuum filtration technique (Technique 11E) may be advisable.

6. Many chemical principles, techniques, and safety are a part of this experiment—make students aware of the complexity of the experiment, but also highlight the rewards of the experiment.

7. Class data are to be collected for the percent acid in the aspirin for this experiment. Describe how the collection of class data is to be done. The analysis of additional data is particularly significant to students who wish to compare their results (yield, % purity, etc.) with other students.

CAUTIONS & DISPOSAL

• Keep the open flames from the boiling water bath away from the organic compounds.

• **Part A.1.** Avoid skin contact with salicylic acid, acetic anhydride, and the conc H_2SO_4.

• The wastes from the synthesis of the aspirin in Part A and the analysis of the aspiring in Part C can be discarded in the sink, followed by a flush with tap water.

• The aspirin synthesized in the laboratory is *not* to be used as a painkiller! The quality control and the purity of the sample prepared in this laboratory is not the same as that required by the FDA.

• The oil bath used for the melting point determination may be as hot as 140°C. Be cautious.

• Request students *not* to throw away the cooking oil used in Part B. It should be returned, when cool, to an appropriately marked container.

TEACHING HINTS

1. **Part A.1.** (Again) emphasize the care of handling the organic compounds.

2. **Part A.3.** Supervise the operation of the vacuum filtration operation.

3. **Part A.4.** The use of "cold" water reduces the amount of aspirin that dissolves in the washing process. Some aspirin may form in the filtrate; this filtrate should be again filtered on another piece of filter paper.

4. **Part A.5.** The recrystallization step produces a purer product; the impurities remain in the supernatant. The recrystallization step is "automatic", no problems are encountered.

5. **Part A.7.** Students sometimes forget about the residual solubility of a product, a factor which is appreciated for the more advanced students.

6. **Part B.1.** Students may need help in filling the capillary melting point tube (Figures 2.2 and 2.3). A melting point apparatus can be substituted for Figure 29.2.

7. **Part B.** The melting point of aspirin is 135°C. The melting points collected by the students are usually less because of impurities (colligative effect). Yields are generally above 90%. Oftentimes yields are reported to be greater than 100%, because the product is not completely dry.

8. **Part C.** The analysis of the aspirin is the same as that for the dry acid in Experiment 28, Part B: a known mass of a dry sample of aspirin is titrated to the phenolphthalein endpoint (see color plate) with the standardized 0.1 M NaOH solution. Appropriate titrating techniques (see Technique 16C) are to be followed.

9. Have students clean the burets thoroughly with several rinses of tap water and a final rinse with deionized water before returning them to the stockroom.

CHEMICALS REQUIRED

salicylic acid	2 g/trial	phenolphthalein dropper bottle
acetic anhydride	5 mL/trial	0.1 M NaOH (standardized) prepared
conc H_2SO_4	0.5 mL	according to the procedure in Experiment 9 or
ethanol	20 mL/trial	by stockroom personnel 100 mL
cooking oil (high temperature bath)	50 mL	95% ethanol 30 mL

SPECIAL EQUIPMENT

fume hood		watchglasses	2
110°C thermometer	1	balance (±0.01 g)	
Büchner funnel, flask, and filter paper	1	mortar and pestle	
capillary melting point tube	1	ring stand and buret clamp	1
360°C thermometer and rubber band	1	ring stand and iron support rings	2
hot plate	1/10 students	Bunsen burner	
125- or 250-mL Erlenmeyer flasks	3	flask clamp	1
weighing paper	3	ice	
50-mL buret and buret brush	1	"Waste Heating Oil" container	

PRELABORATORY ASSIGNMENT

1. An antipyretic is a fever-reducing drug.

2. An analgesic is a painkiller.

3. The active ingredient in aspirin is probably the salicylate ion. Salicylic acid is *not* ingested directly because it causes an upset stomach.

4. 2.61 g acetylsalicylic acid (aspirin)

5. a. 1.79×10^{-3} mol acetylsalicylic acid in the aspirin sample
   b. 0.322 g acetylsalicylic acid in the aspirin sample
   c. 97.3% purity

LABORATORY QUESTIONS

1. The aspirin is undoubtedly impure (colligative property).

2. The aspirin needs to be further dried or recrystallized to remove an excess of impurities.

3. *No.* Since the melting point of aspirin is greater than 100°C, an oil bath must be used.

4. *Lower.* The aspirin isolated after Part A.4 is less pure than that after the recrystallization process in Part A.5 and the separation in Part A.6.

5. *High.* A greater volume of NaOH added in the titration implies a greater quantity of aspirin in the sample.

LABORATORY QUIZ

1. Identify two organic functional groups in an aspirin molecule.
   [Answer: an ester and an organic acid]

2. a. Is aspirin more soluble in ethanol or water?   [Answer: ethanol]
   b. Should ethanol or water be used to wash aspirin crystals?   [Answer: water]

3. What two reactants are used for preparing aspirin?
   [Answer: salicylic acid and acetic anhydride]

4. If the aspirin is not dry before its final mass measurement, is the reported percent yield high or low? Explain.   [Answer: high]

5. A 21.4-mL volume of 0.0987 *M* NaOH titrates a 0.397-g sample of aspirin to the phenolphthalein endpoint. Assuming the reaction of the NaOH and aspirin is a one-to-one mole ratio, what is the percent purity of the aspirin sample? The molar mass of acetylsalicylic acid is 180.2 g/mol.   [Answer: 95.9%]

Molar Solubility, Common-Ion Effect

INTRODUCTION
The solubility constant, K_s, of $Ca(OH)_2$ and its molar solubility in the presence of added Ca^{2+} ion are determined in this experiment. Students have little difficulty in completing the experiment and performing the calculations in a 3 hour laboratory period.

WORK ARRANGEMENT
Individuals.

TIME REQUIREMENT
2.5 hours, *if* the standardized 0.05 M HCl is available.

LECTURE OUTLINE

1. Follow the Instruction Routine outlined in "To the Laboratory Instructor."

2. Discuss the principles of molar solubility and K_s.

3. Use LeChâtelier's principle in discussing the effect of the common-ion, Ca^{2+} in this experiment, on the $Ca(OH)_2$ equilibrium.

4. Students should review (or you should review with students) the titration procedures described in Technique 16C before beginning the experiment.

CAUTION & DISPOSAL

- The test solutions in this experiment can be discarded in the sink, followed by a generous supply of tap water.

TEACHING HINTS

1. The saturated $Ca(OH)_2$ and $Ca(OH)_2/CaCl_2$ solutions can be prepared in the stockroom and then made available to all students. The $Ca(OH)_2$ is a very finely divided precipitate and therefore must be let set to settle before any of the saturated supernatant is withdrawn. When this mixture is disturbed, the $Ca(OH)_2$ solid becomes a part of the supernatant and the results for the experiment are generally "high." Set up some method for distributing the solutions without disturbing the solid $Ca(OH)_2$.

2. **Part A.1.** Ask students why the saturated $Ca(OH)_2$ solutions should be prepared with boiled, deionized water. (The dissolved CO_2 in the tap water neutralizes some of the $Ca(OH)_2$.)

3. **Part A.2.** Inspect the titration apparatus setup and the student's titrating techniques, according to the guidelines in Technique 16C.

4. **Part A.3.** The major source of error in this experiment occurs when the finely divided $Ca(OH)_2$ precipitate is disturbed and then transferred from the stock solution to the receiving flask for the titrant.

5. **Part A.4.** Only 2 to 3 drops of methyl orange indicator are needed. See the color plate for the appearance of the endpoint for methyl orange.

6. Have students clean the burets thoroughly with several rinses of tap water and a final rinse with deionized water before returning them to the stockroom.

7. The K_s of $Ca(OH)_2$ is approximately 5×10^{-6}. The molar solubility of $Ca(OH)_2$ in the presence of the $CaCl_2$ is approximately 4×10^{-3} mol/L. Values are oftentimes high because the saturated solutions becomes disturbed and solid $Ca(OH)_2$ is transferred to the receiving flask in the titration.
For a $K_s = 5 \times 10^{-6}$, the molar solubility of $Ca(OH)_2$ is 1.1×10^{-2} mol/L.

$$0.025 \text{ L} \times \frac{1.1 \times 10^{-2} \text{ mol}}{\text{L}} \times \frac{2 \text{ mol OH}^-}{1 \text{ mol Ca}^{2+}}$$

$$= 5.4 \times 10^{-4} \text{ mol OH}^- = 5.4 \times 10^{-4} \text{ mol H}_3\text{O}^+ \text{ for neutralization.}$$

$$5.4 \times 10^{-4} \text{ mol H}_3\text{O}^+ \times \frac{\text{L}}{0.050 \text{ mol HCl}} = 10.8 \text{ mL of } 0.050 \text{ } M \text{ HCl for neutralization of}$$

the OH$^-$ in a saturated Ca(OH)$_2$ solution.

CHEMICALS REQUIRED	Ca(OH)$_2$ (saturated) prepared 1 week in advance 90 mL 0.05 M HCl (standardized) <100 mL methyl orange indicator dropper bottle Ca(OH)$_2$ (saturated with added CaCl$_2$) prepared 1 week in advance 90 mL

SPECIAL EQUIPMENT

125- or 250-mL Erlenmeyer flasks	3	25-mL pipet and pipet bulb	1
50-mL buret and buret brush	1	ring stand and buret clamp	1

PRELABORATORY ASSIGNMENT

1. a. $K_s = [\text{Ag}^+]^2[\text{CrO}_4^{2-}]$ c. $K_s = [\text{Mg}^{2+}]^3[\text{PO}_4^{3-}]^2$
   b. $K_s = [\text{Ca}^{2+}][\text{CO}_3^{2-}]$

2. a. 1.0×10^{-6} mol CoCO$_3$/L
   b. 5.0×10^{-11} mol CoCO$_3$/L

3. a. $[\text{OH}^-] = 2.9 \times 10^{-5}$ mol/L
   b. $[\text{Cd}^{2+}] = 1.4 \times 10^{-5}$ mol/L
   c. $K_s = 1.2 \times 10^{-14}$

4. a. The OH$^-$ ion from a 25.0 mL aliquot, that is in equilibrium with the Ca(OH)$_2$ solid, is titrated with a standardized 0.050 M HCl solution.
   b. The [Ca^{2+}] is assumed to equal one-half the equilibrium molar concentration of the OH$^-$ ion. See Equation 30.4.

5. a. methyl orange
   b. red (see color plate)
   c. yellow

LABORATORY QUESTIONS

1. The addition of CaCl$_2$ shifts the equilibrium in Equation 30.4 to the left, reducing the molar solubility of Ca(OH)$_2$.

2. a. *Too large.* The transferred Ca(OH)$_2$ is also titrated indicating a larger [OH$^-$], a larger [Ca^{2+}], and a larger solubility of Ca(OH)$_2$. Therefore the K_s is greater.
   b. *Too high.* Because the additional OH$^-$ is also titrated, a larger molar solubility of Ca(OH)$_2$ will be reported.

3. *Too high.* Surpassing the endpoint implies more than actual OH$^-$ ion in solution; the more the [OH$^-$], the larger is the K_s value for Ca(OH)$_2$.

4. *No.* The boiled, distilled water contains no acid or base; therefore, the quantity of HCl required for neutralizing the OH$^-$ ion is unaffected by the amount of boiled, distilled water that is added.

5. *Too low if the tap water is acidic.* Tap water is usually acidic, therefore less 0.05 M HCl is required for neutralization. This indicates that less OH$^-$ ion (from Ca(OH)$_2$) is in solution; the K_s value would be lower than actual.
   Too low if the tap water contains Ca^{2+}. Tap water also contains minerals, most often some Ca^{2+}. Its presence would decrease the solubility of Ca(OH)$_2$ (common-ion effect), and also lower the K_s value of Ca(OH)$_2$.

LABORATORY QUIZ

1. How is the molar solubility of a slightly soluble salt affected by the addition of an ion that is common to the salt equilibrium?   [Answer: decreases]

2. What is the expected pH of a saturated $Cd(OH)_2$ solution? The K_s of $Cd(OH)_2$ is 1.2×10^{-14}.   [Answer: 9.46]

3. A 3.11-mL volume of a standardized 0.0025 M HCl solution titrated 25.0 mL of a saturated $Mg(OH)_2$ solution to the phenolphthalein endpoint. Calculate the K_s of $Mg(OH)_2$.   [Answer: 1.5×10^{-11}]

4. Sketch the shape of a titration curve for the titration of a saturated $Ca(OH)_2$ solution with a standard HCl solution.

Experiment **31**

Galvanic Cells, The Nernst Equation

INTRODUCTION	When two redox couples with different reduction potentials are used to construct a galvanic cell, a spontaneous flow of electrons occurs in the external circuit moving from the anode (–) to the cathode (+). A voltmeter or potentiometer can measure the potential difference between the two redox couples and the direction of electron flow. In this experiment, Part A, a table of redox couples is established relative to the Zn^{2+}(0.1 mol/L)/Zn couple. In Part B, a concentration cell consisting of the Cu^{2+}(0.1 mol/L)/Cu and Cu^{2+}(0.001 mol/L), Cu redox couples is set up to illustrate the dependence of Cu^{2+} concentration on cell potentials and electron flow.

In Part C, the potentials for Cu^{2+} solutions of varying concentrations are measured relative to the Zn^{2+}/Zn redox couple; a plot of E_{cell} versus log $[Cu^{2+}]$ is constructed with the known Cu^{2+} solutions. The E_{cell} of a solution with an unknown $[Cu^{2+}]$ is measured. Using the plotted data, the molar concentration of the Cu^{2+} in the unknown solution is determined. If an expanded scale potentiometer is unavailable, students can "dry lab" this part of the Experimental Procedure, starting with the E_{cell} data for the Zn-Cu cell in Part A. |
| **WORK ARRANGEMENT** | Partners. A sharing of the equipment, skills, and data are valuable to students in this experiment. |
| **TIME REQUIREMENT** | 2.5 hours |
| **LECTURE OUTLINE** | 1. Follow the Instruction Routine outlined in "To the Laboratory Instructor."

2. Review the concept of relative electrode (reduction) potentials. The copper-silver cell described in the Introduction could be used for discussion.

3. Provide an overview of the procedure for setting up a galvanic cell in a 24-well tray and for measuring relative electrode potentials. In this experiment, a salt bridge is used to connect two adjacent wells of a 24-well plate (Figure 31.3). Because the volume of the solutions (and therefore the concentration of the solutes) is small, the potential readings should be made rather quickly.

4. Explain how the Nernst equation (Equation 31.8) is used to measure an unknown concentration in this experiment from plotted data (Equation 31.9). Include in your explanation the use of a graph constructed from known data (Figure 31.2).

5. **Part C.** Advise students as to whether Part C is to be completed as a "wet" or as a "dry" lab. |
| **CAUTIONS & DISPOSAL** | • The salt solutions in the 24-well plate are to be discarded in a "Waste Metal Salts" container.

• Provide a waste container for the metals used as electrodes. |
| **TEACHING HINTS** | 1. The solutions are only 0.1 mol/L because of expense and for a reduction of the ionic activity; the 24-well plate is used to reduce the volume of chemical wastes. The measured redox couples may not equal the standard reduction potentials found in the Standard Reduction Potential Table in your textbook because of the 0.1 mol/L concentrations of ions and ionic activity.

2. **Part A.2.** You may want to assist the student in preparing the salt bridge. |

3. **Part A.3.** Advise students of the proper procedure for checking the polarity of the electrodes connected to the voltmeter. Electrons flow from anode(–) to cathode(+); therefore, if the voltmeter reads positive voltage, the electrode attached to the (+) terminal of the voltmeter is where reduction occurs and that redox couple has the *higher* reduction potential.

4. Theoretically, the following voltages should be recorded:
   Zn-Cu, $E_{cell} \approx 1.10$ V
   Cu-Pb, $E_{cell} \approx 0.47$ V
   Zn-Pb, $E_{cell} \approx 0.63$ V
   Fe-Pb, $E_{cell} \approx 0.31$ V
   Zn-Fe, $E_{cell} \approx 0.32$ V

5. The sum of the Zn-Pb and Cu-Pb cell potentials should equal the Zn-Cu cell potential. The sum of the Zn-Fe and Fe-Pb cell potentials should equal the Zn-Pb cell potential Assuming the reduction potential for Zn(0.1 M)/Zn to be –0.79 V, other reduction potentials are
   Cu^{2+}(0.1 mol/L)/Cu, $E = +0.31$ V
   Pb^{2+}(0.1 mol/L)/Pb, $E = -0.16$ V
   Fe^{2+}(0.1 mol/L)/Fe, $E = -0.47$ V

6. During your movement about the lab, continually question students on each of the following points as the experiment is conducted:
   • Which electrode is the cathode?
   • Which electrode is the anode?
   • In which direction is the flow of electrons?
   • Which electrode is the positive electrode?
   • Which electrode is the negative electrode?
   • At which electrode is oxidation occurring?
   • At which electrode is reduction occurring?
   • In which direction are the cations moving in the cell?
   • In which direction are the anions moving in the cell?

7. **Part B.1.** The cell reaction and the Nernst equation can be used to explain the appearance of a voltage. $E_{cell} \approx 0.089$ V; the copper electrode dipped in the 0.001 M Cu^{2+} solution is the anode.

8. **Part B.2.** The voltage increases with the NH_3 addition to the 0.001 M Cu^{2+} solution because of the reduced free Cu^{2+} concentration. The removal of Cu^{2+}(0.001 mol/L) from the cell reaction, Cu^{2+}(1 mol/L) + Cu(s) → Cu^{2+}(0.001 mol/L) + Cu(s), increases the potential for the reaction to proceed *right*; this shows a higher voltage.

9. **Part C.** Note the introductory paragraph—an expanded scale voltmeter is highly recommended since changes in potential are small. The theoretical voltages are:
   Solution 1:   1.10 V
   Solution 2:   1.04 V
   Solution 3:   0.98 V
   Solution 4:   0.92 V

10. **Part C.1.** Students should use a clean, rinsed pipet for each dilution step.

11. **Part C.4.** Approve the graph of $E_{cell(\text{experimental } and \text{ calculated})}$ versus log [Cu^{2+}].

12. Students should thoroughly clean the electrodes, voltmeter, and all connectors before checking in this equipment.

CHEMICALS REQUIRED	Zn, Pb, Cu, Fe (ungalvanized nail) strips		$0.1\ M$ Cu(NO$_3$)$_2$	4 mL
		1 each	$0.1\ M$ FeSO$_4$	
	$1\ M$ HNO$_3$	10 mL	or $0.1\ M$ Fe(NH$_4$)$_2$(SO$_4$)$_2$	2 mL
	$0.1\ M$ KNO$_3$ (dropper bottle)	2 mL	$1\ M$ CuSO$_4$	5 mL
	$0.1\ M$ Zn(NO$_3$)$_2$	2 mL	$0.001\ M$ CuSO$_4$	3 mL
	$0.1\ M$ Pb(NO$_3$)$_2$	2 mL	$6\ M$ NH$_3$	1 mL

SUGGESTED UNKNOWNS

Redox couples may include (be sure the electrodes are polished)
Ag electrode, $0.1\ M$ AgNO$_3$ solution (expensive!)
Ni electrode, $0.1\ M$ Ni(NO$_3$)$_2$ solution
Mg electrode, $0.1\ M$ Mg(NO$_3$)$_2$ solution, heat to remove dissolved air and cool
Sn electrode, $0.1\ M$ SnCl$_2$ solution

An unknown for Part C can be Solution 1, 2, 3, or 4 or some other concentration ranging from 0.1 to 1×10^{-7} mol/L Cu^{2+}. Keep in mind that a concentration change of one order of magnitude only changes the voltage by 0.03 V.

SPECIAL EQUIPMENT

24-well plate and Beral pipets	1
steel wool	
filter paper	3
0.3 V D.C. voltmeter with a 100 ohm/volt resistance and two connecting wires with alligator clips	1

"Waste Metal Solutions" container
labeled containers for metals

Part C only

expanded scale voltmeter (0 to 2 V or 0 to 1.5 V)	1/student group
100-mL volumetric flasks	3/student group
1-mL pipet and pipet bulb	2/student group

PRELABORATORY ASSIGNMENT

1. a. Cl$_2$ + 2 e$^-$ → 2 Cl$^-$
    b. anode: Pb → Pb^{2+} + 2 e$^-$
        cathode: Cl$_2$ + 2 e$^-$ → 2 Cl$^-$
    c. Pb(s) + Cl$_2$(g) → Pb^{2+}(aq) + 2 Cl$^-$(aq)
    d. $E°_{cell}$ = +1.36 V – (–0.13 V) = +1.49 V

2. $E°_{cell}$ = +0.80 V – (–0.74 V) = +1.54 V

3. a. +1.38 V
    b. [Cr^{3+}] = 1.24 $\times$ 10^{-2} mol/L

4. The sign of the cathode is positive in a galvanic cell. Reduction always occurs at the cathode.

LABORATORY QUESTIONS

1.
Cell	Oxidizing Agent	Reducing Agent
Cu-Zn	Cu^{2+}	Zn
Cu-Pb	Cu^{2+}	Pb
Zn-Pb	Pb^{2+}	Zn
Fe-Pb	Pb^{2+}	Fe
Zn-Fe	Fe^{2+}	Zn

2. Several possibilities include:
    • the electrical resistance of the wire and the voltmeter through which electrons flow
    • the inefficient transfer of electrons from the ions to the electrode (and vice versa)
    • the temperature is *not* at 25°C
    • the concentrations of ions are *not* 1.0 mol/L
    • a correction for the activity of the ions in solution is not included

3. Nonmetals are nonconductors of electricity; the one exception is graphite.

LABORATORY QUIZ
1. Oxidation occurs at the (anode, cathode) and is connected to the (positive, negative) terminal of the voltmeter in a galvanic cell.   [Answer: anode, negative]

2. Electron flow is from the anode to the cathode. In a galvanic cell, electrons flow through the voltmeter from the (positive to negative, negative to positive) terminal when a positive voltage is recorded.   [Answer: negative to positive]

3. State the purpose of the salt bridge in a galvanic cell.
   [Answer: to complete the electrical circuit in the solution portion of the galvanic cell]

4. Given the cell reaction:
   $Cu^{2+}(0.1 \text{ mol/L}) + Zn(s) \rightarrow Cu(s) + Zn^{2+}(0.1 \text{ mol/L})$
   a. What happens to the cell potential if the $[Zn^{2+}]$ is increased to 1.0 mol/L?
   [Answer: E_{cell} decreases]
   b. Does the cell potential increase or decrease if the $[Cu^{2+}]$ is decreased to 0.001 mol/L?   [Answer: E_{cell} decreases]

5. Given:
   $Ag^+ + e^- \rightleftharpoons Ag$ $E° = +0.80 \text{ V}$
   $Pb^{2+} + 2 e^- \rightleftharpoons Pb$ $E° = -0.13 \text{ V}$

   a. Calculate $E°_{cell}$.   [Answer: +1.03 V]
   b. Write the cell reaction.   [Answer: $2 Ag^+(aq) + Pb(s) \rightarrow 2 Ag(s) + Pb^{2+}(aq)$]
   c. What is E_{cell} if $[Ag^+] = 0.0010$ mol/L and $[Pb^{2+}] = 0.10$ mol/L? [Answer = 0.88 V]
   d. If E_{cell} is 0.94 V, what is the $[Ag^+]$, assuming the $[Pb^{2+}]$ remains at 0.10 mol/L?
   [Answer: 9.5×10^{-3} mol/L]

Electrolytic Cells, Faraday's Laws

INTRODUCTION	In this experiment a direct current (dc) power supply causes several electrochemical reactions to occur. Cathode and anode products are detected by testing with litmus paper and by observing color changes and the evolution of gases. In addition, data, that are based on the underlying principles of Faraday's laws, are collected from the electroplating of copper metal to calculate Avogadro's number and the Faraday constant.
WORK ARRANGEMENT	Partners. The interchange of observations and conclusions prove very beneficial in this experiment.
TIME REQUIREMENT	2.5 hours

LECTURE OUTLINE

1. Follow the Instruction Routine outlined in "To the Laboratory Instructor."

2. Define decomposition potential; realize that a calculated decomposition potential does not consider any overpotential, internal resistance of the cell, etc.

3. Students must realize that the oxidation (anode) and reduction (cathode) of water are always possible reactions in the electrolysis of an aqueous solution.
   anode: $2 H_2O(l) \rightarrow O_2(g) + 4 H^+(aq) + 4 e^-$
   cathode: $2 H_2O(l) + 2 e^- \rightarrow H_2(g) + 2 OH^-(aq)$

4. Explain how the measurements for the determination of Avogadro's number and the Faraday constant are conducted in Part B.

CAUTION & DISPOSAL

• Discard all test solutions in a "Waste Salts" container.

TEACHING HINTS

1. **Part A.1.** Check the electrolysis apparatus.

2. **Part A.2.** Observe the technique for the litmus test; make sure the test (Technique 17B) is performed correctly. The anode is the (+) terminal of the D. C. power supply; the cathode is the (–) terminal.

3. **Part A.2.** Close observations of the electrode reactions are necessary. Occasionally gas bubbles appear on an electrode because of the dissolved gases in the solution—heating the solution before electrolysis eliminates this possible observation error.

4. During your movement about the laboratory, continually question students on each of the following points as the experiment is conducted:
   • Which electrode is the cathode?
   • Which electrode is the anode?
   • In which direction is the flow of electrons?
   • Which electrode is the positive electrode?
   • Which electrode is the negative electrode?
   • At which electrode is oxidation occurring?
   • At which electrode is reduction occurring?
   • In which direction are the cations moving in the cell?
   • In which direction are the anions moving in the cell?

5. **Part A.2.** Products of electrolysis:

Solution	Anode	Cathode
NaCl	$2 H_2O(l) \rightarrow$ $O_2(g) + 4 H^+(aq) + 4 e^-$	$2 H_2O(l) + 2 e^- \rightarrow H_2(g) + 2 OH^-(aq)$

NaBr	$2\,Br^-(aq) \rightarrow Br_2(l) + 2\,e^-$		$2\,H_2O(l) + 2\,e^- \rightarrow H_2(g) + 2\,OH^-(aq)$
NaI	$2\,I^-(aq) \rightarrow I_2(s) + 2\,e^-$		$2\,H_2O(l) + 2\,e^- \rightarrow H_2(g) + 2\,OH^-(aq)$
$CuSO_4$	$2\,H_2O(l) \rightarrow$		$Cu^{2+}(aq) + 2\,e^- \rightarrow Cu(s)$
	$O_2(g) + 4\,H^+(aq) + 4\,e^-$		
$CuSO_4$	$Cu(s) \rightarrow Cu^{2+}(aq) + 2\,e^-$		$Cu^{2+}(aq) + 2\,e^- \rightarrow Cu(s)$

6. **Part B.1.** Approve the apparatus before electrolysis begins.

7. **Part B.2** During the electrolysis, the electrodes should *not* be moved as this changes current flow. The values for Avogadro's number and the Faraday constant from the electrolysis are not particularly accurate, especially when the percent error is calculated. This is a result of the primitive equipment that is used for the determinations. But if the experiment is performed carefully, the experimental value for Avogadro's number can be within an order of magnitude of 6.02×10^{23} e⁻/mol e⁻. Better equipment provides a more accurate value for each.

8. An alternative or an additional challenge to Part B is to do calculations for the determination of Avogadro's number and the Faraday constant. For example, a current of 0.5 A passed through a solution for 15 minutes should electroplate

$$15\ min \times \frac{60\ s}{min} \times \frac{0.5\ C}{s} \times \frac{mol\ e^-}{96500\ C} \times \frac{1\ mol\ Cu}{2\ mol\ e^-} \times \frac{63.54\ g\ Cu}{mol\ Cu} = 0.15\ g\ Cu$$

How does this value compare to the mass of copper that was actually electroplated in the experiment?

9. **Part B.3.** Acetone can be used to assist in the drying of the electrodes.

10. Have students clean and rinse (with tap water and deionized water) the U-tube and the carbon and copper electrodes. The dc power supply and the electrical connectors should be free of all solutions before returning them to the stockroom.

CHEMICALS REQUIRED				
	$NaCl(s)$	2 g	Other soluble salt solutions, such as those in	
	$NaBr(s)$	2 g	Expt 31, can be electrolyzed.	50 mL
	$KI(s)$	2 g	1 M HNO_3	20 mL
	0.1 M $CuSO_4$	150 mL	1.0 M $CuSO_4$	75 mL
	graphite electrodes	2	acetone (optional)	10 mL
	copper wire for electrodes	2		

SPECIAL EQUIPMENT			
	dc power supply, complete with wire leads and alligator clips. Any battery or power source with an output of 4 V is satisfactory. A 9-V lantern battery, two to three flashlight batteries in series, or a 9 V transistor battery is sufficient.	U-tube	1
		litmus paper (red/blue)	2
		steel wool	
		balance (±0.001 g)	
		variable resistor (optional)	1
		ammeter (0.2–1 amperes)	1
		"Waste Salts" container	

PRELABORATORY ASSIGNMENT

1. a. $2\,H_2O(l) \rightarrow O_2(g) + 4\,H^+(aq) + 4\,e^-$
   b. $2\,H_2O(l) + 2\,e^- \rightarrow H_2(g) + 2\,OH^-(aq)$

2. The highly refined copper is electroplated at the cathode (–) at left. The blister (impure) copper is placed at the anode (+) at right.

3. a. Cathode reaction is: $2\,H_2O(l) + 2\,e^- \rightarrow H_2(g) + 2\,OH^-(aq)$
   b. Cathode reaction is: $Ni^{2+}(aq) + 2\,e^- \rightarrow Ni(s)$

4. a. anode: $2\,H_2O(l) \rightarrow O_2(g) + 4\,H^+(aq) + 4\,e^-$

   cathode: $Cd^{2+}(aq) + 2\,e^- \rightarrow Cd(s)$

   minimum voltage: 1.219 V

   anode product(s): O_2, H^+

   cathode product(s):   Cd

b. anode: $Cd(s) \rightarrow Cd^{2+}(aq) + 2e^-$

cathode: $Cd^{2+}(aq) + 2e^- \rightarrow Cd(s)$

minimum voltage: 0.00 V

anode product(s): Cd^{2+}

cathode product(s): Cd

5. a. 450 coulombs and 2.81×10^{21} electrons

b. 4.72×10^{-3} mol e^-

c. 5.95×10^{23} e^-/mol e^-

LABORATORY QUESTIONS

1. If lead electrodes are used instead of graphite, the oxidation of lead may occur instead of the oxidation of water. No change in the cathode product(s) is expected.

2. If the solution is highly acidic, H^+ may be reduced instead of water. In both cases H_2 gas is evolved and in both cases the pH increases:

$2 H^+(aq) + 2 e^- \rightarrow H_2(g)$ (lowering of $[H^+]$, pH increases)

$2 H_2O(l) + 2 e^- \rightarrow H_2(g) + 2 OH^-(aq)$ ($[OH^-]$ increases, pH increases)

Therefore, no change in products would be observed in the tests. However, for a highly acidic solution, H^+ would be consumed; in a neutral solution, OH^- would be produced.

3. a. $\text{mol } e^- = \underline{\quad} \text{ g Cu} \times \dfrac{\text{mol Cu}}{63.54 \text{ g Cu}} \times \dfrac{2 \text{ mol } e^-}{\text{mol Cu}}$

$\text{number of electrons} = \dfrac{0.5 \text{ C}}{\text{sec}} \times \underline{\quad} \text{ sec} \times \dfrac{\text{electron}}{1.6 \times 10^{-19}\text{C}}$

b. Difference occurs because of some oxidation of water at the anode.

4. Generally the current flow *decreases* because of the increased resistance of ion flow in solution and changes in the concentration of Cu^{2+} (although, theoretically neither of these factors should affect the current flow).

5. Initially, because of high chloride concentrations, the Cl^- has a higher potential for oxidation than that of water. However as the chloride concentration decreases during the electrolysis, the water assumes the higher potential for oxidation.

LABORATORY QUIZ

1. A copper-zinc mixture can be resolved by dissolving it with nitric acid and selectively reducing the ions at the cathode. Which ion is the first to be reduced? Explain.

$Cu^{2+} + 2 e^- \rightleftharpoons Cu$ $\qquad E^\circ = +0.34$ V

$Zn^{2+} + 2 e^- \rightleftharpoons Zn$ $\qquad E^\circ = -0.76$ V $\qquad\qquad$ [Answer: Cu^{2+}]

2. In the electrolysis of an aqueous sodium sulfate solution, a gas is evolved at the anode and the solution turns blue litmus red. Write the equation for anode reaction.

[Answer: $2 H_2O(l) \rightarrow O_2(g) + 4 H^+(aq) + 4 e^-$]

3. a. In an electrolytic cell (oxidation, reduction) occurs at the anode and its sign is (positive, negative). [Answer: oxidation, positive]

b. Electrons flow from the _____ to the _____. [Answer: anode, cathode]

4. How many moles of electrons are required to electroplate 0.0371 g of nickel from a $NiSO_4$ solution? The molar mass of nickel is 58.3 g/mol. [Answer: 1.26×10^{-3} mol e^-]

5. How many electrons pass through an electrolytic cell if a 0.47 ampere current is applied for 15 minutes? [Answer: 2.64×10^{21} electrons]

6. Electrolysis of molten KI produces potassium metal at the cathode; however, in the electrolysis of an aqueous KI solution, H_2 and OH^- are produced at the cathode. Explain. [Answer: Water has a higher reduction potential than potassium ion.]

Preface to Qualitative Analysis

INTRODUCTION　　Assignment of one or more of the following four experiments is for many students the most challenging, and yet most enjoyable, phase of the general chemistry laboratory. Students often spend hours outside of the regular laboratory period to ensure that the ion(s) in their unknown test solution is correctly identified. We encourage students to set aside "extra" laboratory time during the qualitative analyses, especially when Experiment 36, the General Unknown, is assigned. The development and improvement of laboratory techniques, the practice of making close observations to promote logical conclusions, and the exposure to the solution chemistry of common ions make qualitative analysis a fitting conclusion to any general chemistry program.

Principles used in the isolation and identification of the ions include pH, acid-bases, buffers, oxidation-reduction, the solubility of salts, and complex formation. Continuously discuss these principles with the students as they conduct the experiments so that the procedure does not take on a "cookbook" approach. Students should realize that each step in the procedure has a purpose for separating and/or confirming the presence of an ion. Not understanding the chemistry of the "qual scheme" is the most common cause of carelessness in cation identification.

You have a tremendous responsibility in making qualitative analysis a meaningful experience. At some point in your contact with students, implicate the integrated chemical concepts that are used in the "qual scheme."

LECTURE OUTLINE　　1.　A general overview of qualitative analysis should be given prior to any one of the "qual" experiments. Discuss both the anion and cation qualitative analysis experiments as appropriate.

2.　Remind students that extra laboratory time may be necessary for identifying the ion(s) in and unknown test solution.

3.　Encourage students to follow the chemistry of the separations and confirmatory tests—if they can't, they should ask.

4.　Review the techniques listed in the Introduction for handling test solutions in the qualitative analysis:
　• Measuring and mixing test solutions
　• Testing for complete precipitation
　• Washing a precipitate
　• Heating and cooling solutions
　• Also, review the technique of operating a centrifuge (Technique 11F)

5.　Discuss the construction and use of a flow diagram. Students require assistance in sketching their *first* flow diagram, but soon come to rely on the diagram as an overall guide for identification, especially if the flow diagrams are consistent with respect to notations.

6.　Hints on "How to Effectively do Qual" are presented at the conclusion of the Dry Lab. This should be reviewed with or by students.

7.　Require students to complete the Prelaboratory Assignments *before* the laboratory procedure begins.

8.　Advise students to review the figures closely as they read through the Experimental Procedures.

1. Before and during the analysis, make students aware of the principle(s) being used in the step they are performing: "Why are you adding HNO_3? Is it an acid-base or redox reaction? Is it a displacement reaction?" A continuous barrage of probing questions focuses on many of the chemical principles studied during the year in general chemistry. There is a reason for each step—help the student understand its purpose. Avoid qualitative analysis from being a "cookbook" of experiments.

2. Demonstrate the proper use of a centrifuge.

3. Watch closely the handling of solutions in test tubes.

4. Encourage students to follow a flow diagram (and to make corresponding marginal notes) as they perform the experiment.

5. Each numbered superscript in the Experimental Procedure requires the recording of data on the Report Sheet.

6. Have the students perform tests on their known and unknown test solutions simultaneously as they follow the Experimental Procedure.

7. We grade our unknowns as follows:

$$\frac{\text{number of ions reported } \textit{correctly}}{\text{number of ion present + number of ions reported } \textit{incorrectly}} \times 100$$

8. If "unexpected" observations appear at any point in the "qual scheme", consult any qualitative analysis text. Occasionally interferences from other ions occur, especially when the test solutions become contaminated.

1. a. 3 mL
    b. dropping pipet
    c. stirring rod
    d. centrifuge
    e. supernatant
    f. 20 drops
    g. 20–40 seconds
    h. the formation of a precipitate
    i. supernatant (or decantate)
    j. confirmatory identification of the ion
    k. the presence of soluble ions
    l. tapping the side of the test tube with the "pinky" finger
    m. using a hot water bath

2. See Technique 11F.

3. Add the wash liquid to the precipitate, mix thoroughly with a stirring rod, centrifuge, and decant.

4. First, add the precipitating reagent and centrifuge. Next, add 1–2 drops of the precipitating reagent; if more precipitate forms, centrifuge again. Repeat the reagent/centrifuge cycle until no further precipitation occurs.

5. The elements to be identified in the Periodic Table are:

Common Anions	Qual I Cations	Qual II Cations	Qual III Cations
S for SO_4^{2-} and S^{2-}	Na for Na^+	Mn for Mn^{2+}	Mg for Mg^{2+}
P for PO_4^{3-}	K for K^+	Ni for Ni^{2+}	Ca for Ca^{2+}
C for CO_3^{2-}	N for NH_4^+	Fe for $Fe^{3+(2+)}$	Ba for Ba^{2+}
Cl for Cl^-	Ag for Ag^+	Al for Al^{3+}	
Br for Br^-	Cu for Cu^{2+}	Zn for Zn^{2+}	
I for I^-	Bi for Bi^{3+}		
N for NO_3^-			

Common Anions

INTRODUCTION

This experiment is the first of a very important set of qualitative analysis experiments in this manual—the cation analyses are outlined in Experiments 34 through 36.

A thorough understanding of the many chemical principles that are illustrated in qualitative analysis changes "qual" experiments from a "cookbook" operation into a very valuable learning experience. A student's attitude toward "qual" is reflected by the instructor's enthusiasm. A constant barrage of questions and explanations (on a one-to-one basis) during the laboratory period makes the experience more meaningful.

WORK ARRANGEMENT

Partners on the known test solution; individuals on an unknown test solution.

TIME REQUIREMENT

3 hours

LECTURE OUTLINE

1. Follow the Instruction Routine outlined in "To the Laboratory Instructor."

2. Discuss the various techniques common to qualitative analysis, described in Dry Lab 5:
   • measuring and mixing test solutions
   • testing for complete precipitation
   • washing a precipitate
   • heating and cooling solutions

3. Review the completed anion flow diagram in the manual. Use the anion flow diagram to account for the symbols and the organization of a flow diagram. Have students refer to it often as they work their way through the experimental procedure.

4. Include Equations 33.1–13 in your review of the anion flow diagram.

5. Students use the centrifuge often in this and subsequent qualitative analysis experiments. You may want to review the proper technique (Technique 11F) for operating a centrifuge. Students need to be made aware of the size of the centrifuge tubes that fit the centrifuge; ideally 75-mm test tubes will fit, but you will need to check the centrifuge in your laboratories.

CAUTIONS & DISPOSAL

• Acids and bases: 6 M NH_3 (Parts A.1 and H.2), 6 M HNO_3 (throughout the experiment), and conc H_2SO_4 (Part I.2) are used in the experiment. Be aware of the correct procedure (Laboratory Safety section) for cleaning up acid/base spills.

• Acidic test solutions can be discarded in the "Waste Acids" container.

• Dispose of the toluene test samples in the "Waste Organics" container.

TEACHING HINTS

1. Each numbered superscript in the Experimental Procedure requires an observation to be recorded on the Report Sheet.

2. Have the students perform tests on a sample containing known anions and a sample containing their "unknown" anions simultaneously as they follow the Experimental Procedure. Students should *not* discard the compound that confirms the presence of an anion; it should be kept for comparing observations.

3. **Part D.1.** Before beginning the test for CO_3^{2-} ion, the PO_4^{3-} the SO_4^{2-} ions are to be removed.

4. Make sure students use proper techniques in completing the experiment—the icons are for their benefit in properly completing the procedures.

5. Data for the tests for the various steps of the Experimental Procedure.

Procedure Number and Ion	Test Reagent or Technique	Observation (Color or General Appearance)		Chemical(s) Responsible for Observation	Equation(s) for Observed Reaction
#1	$NH_3/$ $Ba(NO_3)_2$	ppt	white	$BaSO_4$, $BaCO_3$, $BaSO_4$	$Ba^{2+} + (SO_4^{2-}, CO_3^{2-}, PO_4^{3-}) \rightarrow$ $BaSO_4$, $BaCO_3$, $BaSO_4$
#2		spnt	colorless	S^{2-}, Cl^-, Br^-, I^-, NO_3^-	
#3 SO_4^{2-}	HNO_3	ppt	white	$BaSO_4$	33.1
#4 PO_4^{3-}	$(NH_4)_2MoO_4$	ppt	yellow	$(NH_4)_3PO_4(MoO_3)_{12}$	33.3
#5 CO_3^{2-}	$HNO_3/$ $Ca(OH)_2$	ppt	white	$CaCO_3$	33.6
#6 S^{2-}	$HNO_3/$ $Cu(NO_3)_2$	ppt	black	CuS	33.7
#7 I^-	$Fe(NO_3)_3/$ toluene	color	purple	I_2	33.8
#8 Br^-	$HNO_3/$ $KMnO_4$	color	brown	Br_2	33.9
#9 Cl^-	$AgNO_3$	ppt	white	$AgCl$	33.10
#10	NH_3	spnt	colorless	$[Ag(NH_3)_2]^+$	33.11
#11	Ag_2SO_4	ppt	white	$AgCl$, $AgBr$, AgI, Ag_2SO_4, Ag_3PO_4, Ag_2CO_3, Ag_2S	$Ag^+ + (Cl^-, Br^-, I^-, SO_4^{2-}, PO_4^{3-}, CO_3^{2-}, S^{2-}) \rightarrow AgCl$, $AgBr$, AgI, Ag_3PO_4, Ag_2CO_3, Ag_2S
#12 NO_3^-	conc H_2SO_4	color	brown	$FeNO^{2+}$	33.13

CHEMICALS REQUIRED

6 M NH_3	3 mL	0.01 M $AgNO_3$ (dropper bottle)	1 mL
1 M $Ba(NO_3)_2$ (dropper bottle)	1 mL	toluene (dropper bottle)	1 mL
6 M HNO_3	6 mL	3 M NaOH (dropper bottle)	0.5 mL
0.5 M $(NH_4)_2MoO_4$	2 mL	0.04 M (satd) Ag_2SO_4	1 mL
satd $Ca(OH)_2$, limewater	0.5 mL	3 M H_2SO_4(dropper bottle)	0.5 mL
1 M $Cu(NO_3)_2$ (dropper bottle)	1 mL	satd $FeSO_4$	1 mL
0.2 M $Fe(NO_3)_3$ (dropper bottle)	1 mL	conc H_2SO_4 (dropper bottle)	1 mL
0.1 M $KMnO_4$ (dropper bottle)	1 mL		

SUGGESTED UNKNOWNS

0.2 M Na_2SO_4	10 mL	0.2 M NaBr	10 mL
0.2 M Na_2CO_3	10 mL	0.2 M NaI	10 mL
0.2 M Na_3PO_4	10 mL	0.2 M Na_2S	10 mL
0.2 M NaCl	10 mL	0.2 M $NaNO_3$	10 mL

Combine 2 mL of each solution for a *known* test solution.
Prepare the *unknown* test solution by substituting 2 mL of deionized water for each anion *not* present.

SPECIAL EQUIPMENT

centrifuge	1/5 students	ice
75-mm test tubes	extra	Bunsen burner
test tube holder		ring stand and iron support rings
litmus (red and blue)		"Waste Organics" container
dropping (Beral) pipets		"Waste Acid" container

1. $6\ M\ HNO_3$ dissolves $Ba_3(PO_4)_2$ but not $BaSO_4$.

2. Fe^{3+} oxidizes I^- to I_2; after the oxidation of the I^-, MnO_4^- oxidizes Br^- to Br_2.

3. $Ca(NO_3)_2$. $CaCO_3$ is insoluble, $BaCl_2$ is soluble.

4. *Twice.* Once for Parts B and C and a second time for Part D.

5.
   a. yellow precipitate $(NH_4)_3PO_4(MoO_3)_{12}$
   b. white precipitate $BaSO_4$
   c. black precipitate CuS
   d. brown solution in toluene Br_2
   e. white precipitate $AgCl$

LABORATORY QUESTIONS

1. $HCl(aq)$ causes the precipitation of $AgCl$, whereas Ag^+ does not react with the $HNO_3(aq)$.

2. Toluene is a nonpolar molecule as is Br_2 and I_2, the oxidation products of Br^- and I^-; a greater solubility occurs between the halogens and toluene than between the halogens and water.

3. Ag_2SO_4 is used to precipitate all interfering ions of the nitrate ion test.

*4. Most oxalates are insoluble, including BaC_2O_4. Therefore, it would appear in the precipitate of Part A.2.

LABORATORY QUIZ

1. Which anions precipitate in the presence of Ba^{2+}?
   a. Cl^- d. CO_3^{2-}
   b. PO_4^{3-} e. NO_3^-
   c. S^{2-} f. Br^- [Answer: PO_4^{3-} and CO_3^{2-}]

2. How are phosphate ions separated from sulfate ions in an aqueous solution?
   [Answer: precipitate both with Ba^{2+} and then add $HNO_3(aq)$; $Ba_3(PO_4)_2$ dissolves, $BaSO_4$ does not.]

3. Which anion(s) do *not* precipitate in the presence of Ag^+?
   a. SO_4^{2-} d. I^-
   b. NO_3^- e. CO_3^{2-}
   c. S^{2-} f. PO_4^{3-} [Answer: all precipitate as the silver salt, except NO_3^-]

4. How are bromide ions identified in the presence of iodide ions?
   [Answer: See answer to Prelab Question 2]

5. Tell what happens to each of the following salts when nitric acid is added to it:
   a. $BaCO_3$ [Answer: precipitate dissolves and a gas is evolved]
   b. $Ba_3(PO_4)_3$ [Answer: precipitate dissolves]
   c. $BaSO_4$ [Answer: nothing]

6. What is the color of each?
   a. $AgCl$ d. $BaSO_4$
   b. $(NH_4)_3PO_4(MoO_3)_{12}$ e. Br_2
   c. PbS f. I_2

Qual I. Na$^+$, K$^+$, NH$_4^+$, Ag$^+$, Cu^{2+}, Bi^{3+}

INTRODUCTION

Fourteen cations are separated and identified in the next three experiments:

Experiment 34: Qual I.   Na$^+$, K$^+$, NH$_4^+$, Ag$^+$, Cu^{2+}, Bi^{3+}
Experiment 35: Qual II.   Mn^{2+}, Ni^{2+}, Fe^{3+}(Fe^{2+}), Al^{3+}, Zn^{2+}
Experiment 36: Qual III. Mg^{2+}, Ca^{2+}, Ba^{2+} and General Unknown Examination

The experiments are written so that a test solution containing an unknown number of cations from several groups can be assigned for the General Unknown in Experiment 36. If you plan to assign a General Unknown, announce your plans to the class at this time. Experiment 36 includes a suggested procedure for analyzing a General Unknown test solution containing any number of cations from the three groups.

Students are advised to analyze each "qual" group separately. A known sample containing all of the cations from the respective group should be analyzed concurrently with a sample containing the unknown cation(s) in a test solution.

An analysis of the cations from Qual I requires most of the time of a scheduled 3 hour laboratory. Students require time to set up test solutions and learn the technique for operating the centrifuge (Technique 11F).

The chemical principles used in this experiment include ionic equilibria, precipitation, acid-base control, and complex formation. Review them with students as the situation arises in the analysis.

WORK ARRANGEMENT

Partners on a test solution containing all of the cations; individuals on a test solution containing an unknown number of the Qual I cations.

TIME REQUIREMENT

3 hours

LECTURE OUTLINE

1. Follow the Instruction Routine outlined in "To the Laboratory Instructor."

2. Discuss the various techniques common to qualitative analysis, described in Dry Lab 5:
    • measuring and mixing test solutions
    • testing for complete precipitation
    • washing a precipitate
    • heating and cooling solutions

3. In this first "cation qual" laboratory session, discuss with the aid of a flow diagram:
    • the three qual groups for the cations
    • how the three qual groups are separated
    • the interference cause by improper separation techniques within and between the three qual groups
    • the semimicro techniques used for separation and identification

4. Explain the effect of H$_3$O$^+$ on the sulfide concentration in the equilibrium, H$_2$S(aq) $\rightleftharpoons$ 2 H$^+$(aq) + S^{2-}(aq); a high [H$_3$O$^+$] pushes the equilibrium to the *left*, reducing the [S^{2-}]. With a low [S^{2-}] only the cation sulfides with *small* K_s values precipitate; therefore, pH control separates the Qual I cation sulfides, which all have very small K_s values, from the Qual II cation sulfides, which have larger K_s values.

5. Require the students to complete the Qual I flow diagram on the Prelaboratory Assignment (Question 8) *before* they begin the Experimental Procedure.

6. Discuss the separation and identification of the Qual I cations, using the flow diagram on the Prelaboratory Assignment and Equations 34.1–12.

<table>
<tr><td>CAUTIONS
& DISPOSAL</td><td>•</td><td>Acids and Bases. $6\ M$ HCl, $6\ M$ NaOH, $6\ M$ and conc NH_3, $6\ M$ HNO_3 and $6\ M$ CH_3COOH are all used in the experiment.
Be aware of the correct procedure for cleaning up acid/base spills.</td></tr>
<tr><td></td><td>•</td><td>Advise students to discard their test solutions in the appropriately marked "Waste Metal Salts" container.</td></tr>
</table>

TEACHING HINTS If time is a factor in completing this experiment in the allotted time, assign a single cation for unknown solution of the Qual I cations.

1. **Part A.** Supervise students who are performing the flame test for the first time; you may wish to demonstrate the technique for them. If the wire is not properly cleaned, Na^+ may be mistakenly reported as being present as Na^+ is a "universal" impurity.

2. **Part B.** For any flame test it is always advisable to conduct the test on a solution known to contain, for example, K^+, and then compare its appearance with that of the test solution.

3. **Part C.** The nose is also a good detector of ammonia, although not recommended in this experiment.

4. **Part E.** Omit Part E if cations from qual groups Qual II and Qual III are absent.

5. **Part F.** Generally the second confirmatory test for Cu^{2+} is unnecessary, but is convincing.

6. **Part G.** The sodium stannite must be prepared just prior to its use—it should *not* be prepared in the stockroom.

7. If a general unknown is to be assigned (Experiment 36), give each student the five possible cations that may be in his/her unknown test solution. At least two of the three "qual" groups should be represented in the general unknown test solution. Also encourage them to read the section of Experiment 36 that discusses the General Unknown.

8. Data of the tests for the various steps of the Experimental Procedure.

Procedure Number and Ion	Test Reagent or Technique	Observation (Color or General Appearance)		Chemical(s) Responsible for Observation	Equation(s) for Observed Reaction
#1 Na^+	flame		yellow	—	—
#2 K^+	flame		lavender	—	—
#3 NH_4^+	NaOH	gas	gas	NH_3	34.4
#4 Ag^+	HCl	ppt	white	AgCl	$Ag^+ + Cl^- \rightarrow AgCl$
#5	NH_3	spnt	colorless	$[Ag(NH_3)_2]^+$	34.6
#6	HNO_3	ppt	white	AgCl	34.7 (shift to left)
#7	H_2S/H^+	ppt	black	CuS, Bi_2S_3	$Cu^{2+} + S^{2-} \rightarrow CuS$ $2\ Bi^{3+} + 3\ S^{2-} \rightarrow Bi_2S_3$
#8	HNO_3	spnt	colorless (blue)	Cu^{2+}, Bi^{3+}	34.8 and similar equation for Bi_2S_3

#9 Cu^{2+}	NH_3	spnt	deep blue	$[Cu(NH_3)_4]^{2+}$	34.10
#10		ppt	white	$Bi(OH)_3$	34.9
#11	$K_4[Fe(CN)_6]$	ppt	red-brown	$Cu_2[Fe(CN)_6]$	34.11
#12 Bi^{3+}	$NaSn(OH)_4$	ppt	black	Bi	34.12

CHEMICALS REQUIRED

$CaO(s)$ or $Ca(OH)_2(s)$	pinch	$6\ M\ HNO_3$ (dropper bottle)	3 mL
$(NH_4)_2CO_3(s)$	pinch	$1\ M\ CH_3CSNH_2$ (thioacetamide)	2 mL
$6\ M\ HCl$ (dropper bottle)	1 mL	$0.1\ M\ HCl$	2 mL
$0.2\ M\ NaCl$ (dropper bottle)	0.5 mL	conc NH_3 (dropper bottle)	1 mL
$0.2\ M\ KCl$ (dropper bottle)	0.5 mL	$6\ M\ CH_3COOH$ (dropper bottle)	1 mL
$6\ M\ NaOH$ (dropper bottle)	2 mL	$0.2\ M\ K_4[Fe(CN)_6]$ (dropper bottle)	1 mL
$6\ M\ NH_3$ (dropper bottle)	2 mL	$1\ M\ SnCl_2$	0.5 mL

SUGGESTED UNKNOWNS

$0.25\ M\ NaNO_3$	10 mL	$0.1\ M\ AgNO_3$	10 mL
$0.25\ M\ KNO_3$	10 mL	$0.1\ M\ Cu(NO_3)_2$	10 mL
$0.5\ M\ NH_4NO_3$	10 mL	$0.1\ M\ Bi(NO_3)_3$	10 mL

Combine 2 mL of each solution for a *known* test solution.
Prepare the *unknown* test solution by substituting 2 mL of deionized water for each anion *not* present.

SPECIAL EQUIPMENT

flame test wire
evaporating dish
cobalt blue glass
litmus (red and blue)
watchglass

centrifuge
Bunsen burner
ring stand and iron support rings
wire gauze
"Waste Metal Salts" container

2

PRELABORATORY ASSIGNMENT

1. Because many cations exhibit a color in a flame test, a single flame test on a solution containing several cations will produce a flame with a mixed set of colors.

2. If no precipitate forms when HCl is added to the solution in Part D, then Ag^+ ion is absent from the sample.

3.
Equation 34.2	b, acid-base
Equation 34.3	b, acid-base
Equation 34.6	d, complex formation
Equation 34.8	a, oxidation-reduction
Equation 34.9	b, acid-base
Equation 34.10	d, complex formation
Equation 34.11	d, complex formation

4. Ammonia forms the soluble $[Cu(NH_3)_4]^{2+}$ complex with Cu^{2+}, but forms the $Bi(OH)_3$ precipitate with Bi^{3+}.

5. a. $[Cu(NH_3)_4]^{2+}$, deep blue
   b. bismuth, black
   c. CuS and Bi_2S_3, black

6. $[S^{2-}] = \dfrac{1.1 \times 10^{-21}[0.1]}{[0.316]^2} = 1.1 \times 10^{-21}$ mol/L

7. Flow diagram for Qual I Cations

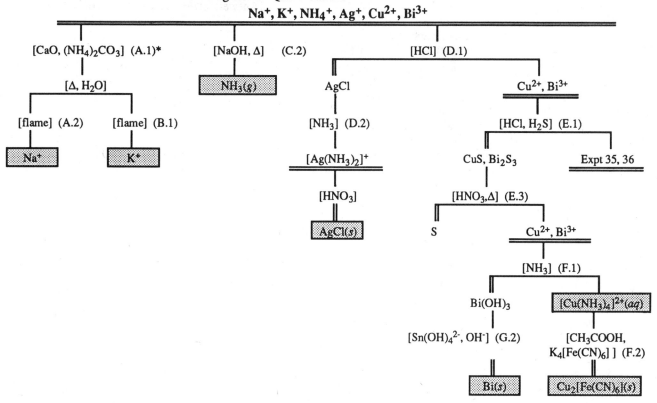

$$Na^+, K^+, NH_4^+, Ag^+, Cu^{2+}, Bi^{3+}$$

[CaO, $(NH_4)_2CO_3$] (A.1)*

[Δ, H_2O]

[flame] (A.2)   [flame] (B.1)

Na⁺   K⁺

[NaOH, Δ]   (C.2)

$NH_3(g)$

[HCl] (D.1)

AgCl Cu^{2+}, Bi^{3+}

[NH_3] (D.2)   [HCl, H_2S] (E.1)

$[Ag(NH_3)_2]^+$ CuS, Bi_2S_3 Expt 35, 36

[HNO_3]   [HNO_3, Δ] (E.3)

AgCl(s)   S Cu^{2+}, Bi^{3+}

[NH_3] (F.1)

$Bi(OH)_3$ $[Cu(NH_3)_4]^{2+}(aq)$

[$Sn(OH)_4^{2-}$, OH^-] (G.2)   [CH_3COOH, $K_4[Fe(CN)_6]$] (F.2)

Bi(s) $Cu_2[Fe(CN)_6](s)$

LABORATORY QUESTIONS

1. a. white   d. deep blue
   b. black   e. colorless
   c. white

2. a. If Ag^+ is present in the test solution, no precipitate will form with the addition of HNO_3; $AgNO_3$ is soluble.
   b. If NaOH is added to insoluble AgCl, insoluble Ag_2O forms in its place because of its lower molar solubility than AgCl; when NH_3 is added to insoluble AgCl, the soluble complex, $[Ag(NH_3)_2]^+$, forms.

3. a. If NaOH instead of NH_3 is added to a test solution containing both Cu^{2+} and Bi^{3+}, both would precipitate as the hydroxides, thereby *not* resulting in their separation.
   b. The white precipitate is $Bi(OH)_3$; conc HCl would cause the dissolution of the $Bi(OH)_3$ precipitate.

4. a. HCl(aq) causes the precipitation of Ag^+, but not Cu^{2+}
   b. HCl(aq) causes the precipitation of Ag^+, but not Na^+
   c. NaOH causes NH_4^+ to be evolved as $NH_3(g)$ and Cu^{2+} to precipitate as $Cu(OH)_2$.

LABORATORY QUIZ

1. How can Bi^{3+} ions be separated from Ag^+ ions?   [Answer: addition of HCl(aq)]

2. When $NH_3(aq)$ is added to a white precipitate, the precipitate dissolves. What is the precipitate?   [Answer: AgCl]

3. a. From the equilibrium,
   $$H_2S(aq) + 2 H_2O(l) \rightleftharpoons 2 H_3O^+(aq) + S^{2-}(aq),$$
   is the [S^{2-}] high or low at a low pH? Explain.   [Answer: low]
   b. From a solution that is 0.1 M Cu^{2+} and 0.1 M Pb^{2+}, which ion precipitates at the *lower* pH from a saturated H_2S solution?
   K_s of CuS = 8.5 x 10^{-36}, K_s of PbS = 3.7 x 10^{-19}, a concentrated H_2S solution has a concentration of 0.1 mol/L.   [Answer: CuS]

4. How is Cu^{2+} separated from Bi^{3+} in the Qual I separation of cations?

    [Answer: addition of an excess of $NH_3(aq)$)]

5. In the reaction,

    $2 Bi(OH)_3(s) + 3 Sn(OH)_3^-(aq) + 3 OH^-(aq) \rightarrow 2 Bi(s) + 2 Sn(OH)_6^{2-}(aq)$,

    what is the oxidizing agent?   [Answer: $Bi(OH)_3$]

6. Write the formulas for:
    a. hydrochloric acid
    b. sodium hydroxide
    c. nitric acid
    d. copper(II) chloride   [Answer: $HCl(aq)$; $NaOH$, $HNO_3(aq)$, $CuCl_2$]

Qual II. Mn^{2+}, Ni^{2+}, $Fe^{3+}(Fe^{2+})$, Al^{3+}, Zn^{2+}

INTRODUCTION

The identifications of the Qual II cations are as challenging at those of Qual I. Many separations and test reagents are used in the analysis and, therefore, advanced preparation for the laboratory is necessary for a successful completion and understanding of the experiment in a reasonable time.

The Qual II analysis is the most colorful of the three cation "qual" groups—the confirmatory tests are purple, red-orange, brick-red, red, and green colors.

The chemical principles used in this experiment are ionic equilibrium, pH control, amphoterism, precipitation, complex-ion formation, oxidation-reduction, and adsorption. Have students cite these principles in the experiment.

WORK ARRANGEMENT

Partners can share observations and discussion on a test solution containing all of the cations; individuals on a test solution containing an unknown number of the Qual II cations.

TIME REQUIREMENT

3 hours

LECTURE OUTLINE

1. Follow the Instruction Routine outlined in "To the Laboratory Instructor."

2. Review the separation and identification of Qual II cations.

3. Discuss how the lower $[H_3O^+]$ increases the $[S^{2-}]$ in the H_2S equilibrium in Equation 35.1. The lower $[H^+]$ (lower than that of Experiment 34), produces a greater $[S^{2-}]$ which is necessary to precipitate the Qual II cations. The equilibrium, $H_2S(aq) + 2\,H_2O(l) \rightleftharpoons 2\,H_3O^+(aq) + S^{2-}(aq)$, shifts *right*, producing a higher S^{2-} concentration.

4. Students should complete the flow diagram on the Prelaboratory Assignment before beginning the experiment.

5. Discuss the completed Qual II flow diagram in conjunction with Equations 35.3–18.

CAUTIONS & DISPOSAL

- Acids and Bases. $6\,M$ and conc NH_3, $6\,M$ and HCl (Part A.2), $6\,M$ and conc HNO_3, $6\,M$ NaOH, and $6\,M$ HCl are all used in the experiment.
  Be aware of the correct procedure for cleaning up acid/base spills; have sodium bicarbonate available in the laboratory

- Advise students to properly dispose of the test solutions and precipitates in the "Waste Metal Salts" container.

TEACHING HINTS

1. If a general unknown is *not* assigned, students may proceed directly to Part A of the Experimental Procedure.

2. **Part A.2.** No more than 2 mL of conc HNO_3 should be added to the moist residue. A fume hood or an improvised hood can be used to remove HNO_3 fumes.

3. **Part F.2.** The Al^{3+} test is the most troublesome in this experiment. The red aluminon dye adsorbs to the gelatinous $Al(OH)_3$ forming a red precipitate, leaving the solution nearly colorless. The presence of a red solution is *not* confirmatory for Al^{3+}.

4. Students should use the proper techniques for washing precipitates, testing for pH with litmus, and adding conc HNO_3.

5. Data of the tests for the various steps of the Experimental Procedure.

Procedure Number and Ion	Test Reagent or Technique	Observation (Color or General Appearance)		Chemical(s) Responsible for Observation	Equation(s) for Observed Reaction
#1	H_2S/NH_3	ppt	black	NiS, FeS, ZnS, MnS, $Al(OH)_3$	$(Ni^{2+}, Fe^{2+}, Zn^{2+}, Mn^{2+}) + S^{2-} \rightarrow$ NiS, FeS, ZnS, MnS $Al^{3+} + 3\,OH^- \rightarrow Al(OH)_3$
#2	HCl/HNO_3	spnt	amber	$Ni^{2+}, Fe^{3+}, Zn^{2+},$ Mn^{2+}, Al^{3+}	35.3, 35.4, $Al(OH)_3 + 3\,H^+ \rightarrow Al^{3+} + 3\,H_2O$
#3	OH^-	ppt	amber, gel	$Fe(OH)_3, Ni(OH)_2,$ $Mn(OH)_2$	$Mn^{2+} + 2\,OH^- \rightarrow Mn(OH)_2$ $Ni^{2+} + 2\,OH^- \rightarrow Ni(OH)_2$ $Fe^{3+} + 3\,OH^- \rightarrow Fe(OH)_3$
#4		spnt	colorless	$Zn(OH)_4^{2-}, Al(OH)_4^-$	35.5, 35.6
#5	HNO_3	spnt	amber	$Fe^{3+}, Ni^{2+}, Mn^{2+}$	$Mn(OH)_2 + 2\,H^+ \rightarrow Mn^{2+} + 2\,H_2O$ $Ni(OH)_2 + 2\,H^+ \rightarrow Ni^{2+} + 2\,H_2O$ $Fe(OH)_3 + 3\,H^+ \rightarrow Fe^{3+} + 3\,H_2O$
#6 Mn^{2+}	$NaBiO_3$	spnt	violet	MnO_4^-	35.7
#7 $Fe^{3+(2+)}$	NH_3	ppt	red-brown	$Fe(OH)_3$	35.8
#8		spnt	blue	$[Ni(NH_3)_6]^{2+}$	35.9
#9	NH_4SCN	spnt	red-orange	$[FeNCS]^{2+}$	35.10
#10 Ni^{2+}	H_2DMG	ppt	pink/ brick red	$Ni(HDMG)_2$	35.11
#11 Al^{3+}	HNO_3/NH_3	ppt	white	$Al(OH)_3$	35.12, 35.13
#12		spnt	colorless	$[Zn(NH_3)_4]^{2+}$	35.14, 35.15
#13	HNO_3, aluminon, NH_3	ppt	red, gel	$Al(OH)_3 \cdot$ aluminon	35.16
#14 Zn^{2+}	$HCl/$ $K_4[Fe(CN)_6]$	ppt	light green	$K_2Zn_3[Fe(CN)_6]_2$	35.17, 35.18

6. If a general unknown is assigned (Experiment 36), require students to design their personal flow diagrams for their assigned cations by the next laboratory period. If each student's list of five possible cations was not identified in the previous laboratory period, it must be done *now*.

CHEMICALS REQUIRED

2 M NH_4Cl	3 mL
6 M NH_3 (dropper bottle)	3 mL
1 M CH_3CSCH_3 (thioacetamide)	1 mL
6 M HCl (dropper bottle)	3 mL
6 M HNO_3 (dropper bottle)	3 mL
conc HNO_3 (dropper bottle)	4 mL

6 M NaOH (dropper bottle)	4 mL
$NaBiO_3(s)$	0.1 g
conc NH_3 (dropper bottle)	1 mL
0.1 M NH_4SCN (dropper bottle)	1 mL
1% dimethylglyoxime (dropper bottle)	0.5 mL
aluminon reagent (dropper bottle)	0.5 mL
0.2 M $K_4[Fe(CN)_6]$ (dropper bottle)	0.5 mL

SUGGESTED UNKNOWNS	0.1 M Fe(NO$_3$)$_3$	10 mL	0.1 M Al(NO$_3$)$_3$	10 mL
	0.1 M Ni(NO$_3$)$_2$	10 mL	0.1 M Mn(NO$_3$)$_2$	
	0.1 M Zn(NO$_3$)$_2$	10 mL	or 0.1 M MnCl$_2$	10 mL

Combine 2 mL of each solution for a *known* test solution.
Prepare the *unknown* test solution by substituting 2 mL of deionized water for each anion *not* present.

SPECIAL EQUIPMENT	centrifuge		ring stand iron rings	2
	75-mm test tubes	extra	Bunsen burner	
	evaporating dish		"Waste Metal Salts" container	
	litmus (red and blue)			

PRELABORATORY ASSIGNMENT

1.
Equation	Mass Action Expression	K_s Value	Color
MnS(s) $\rightleftharpoons$ Mn^{2+}(aq) + S^{2-}(aq)	K_s = [Mn^{2+}][S^{2-}]	7 x 10^{-16}	pink
NiS(s) $\rightleftharpoons$ Ni^{2+}(aq) + S^{2-}(aq)	K_s = [Ni^{2+}][S^{2-}]	2 x 10^{-21}	black
FeS(s) $\rightleftharpoons$ Fe^{2+}(aq) + S^{2-}(aq)	K_s = [Fe^{2+}][S^{2-}]	3.7 x 10^{-19}	black
ZnS(s) $\rightleftharpoons$ Zn^{2+}(aq) + S^{2-}(aq)	K_s = [Zn^{2+}][S^{2-}]	1.2 x 10^{-23}	white
Al(OH)$_3$(s) $\rightleftharpoons$ Al^{3+}(aq) + 3 OH$^-$(aq)	K_s = [Al^{3+}][OH$^-$]^3	2 x 10^{-33}	white

Molar solubility of MnS = 2.6 x 10^{-8} mol/L, most soluble
Molar solubility of NiS = 4.5 x 10^{-11} mol/L
Molar solubility of FeS = 6.1 x 10^{-10} mol/L
Molar solubility of ZnS = 3.5 x 10^{-12} mol/L, least soluble
Molar solubility of Al(OH)$_3$ = 2.9 x 10^{-9} mol/L

2. a.   ZnS(s) + 2 H$^+$(aq) $\rightarrow$ Zn^{2+}(aq) + H$_2$S(aq)
       Al(OH)$_3$(s) + 3 H$^+$(aq) $\rightarrow$ Al^{3+}(aq) + 3 H$_2$O(l)
   b.   Fe^{3+}(aq) + 3 NH$_3$(aq) + 3 H$_2$O(l) $\rightarrow$ Fe(OH)$_3$(s) + 3 NH$_4$$^+$(aq)
       Ni^{2+}(aq) + 6 NH$_3$(aq) $\rightarrow$ [Ni(NH$_3$)$_6$]$^{2+}$(aq)
   c.   Zn^{2+}(aq) + 4 NH$_3$(aq) $\rightarrow$ [Zn(NH$_3$)$_4$]$^{2+}$(aq)

3. Amphoterism means that the compounds are capable of reacting as an acid or base. Al(OH)$_3$ and Zn(OH)$_2$ may react (as a base) with H$_3$O$^+$ to produce Al^{3+} and Zn^{2+} or they may react (as an acid) with OH$^-$ to produce Al(OH)$_4$$^-$ and Zn(OH)$_4$$^{2-}$.

4. a.   blue
   b.   blood red
   c.   pink to brick red
   d.   deep purple

5. Flow diagram for Qual II cations:

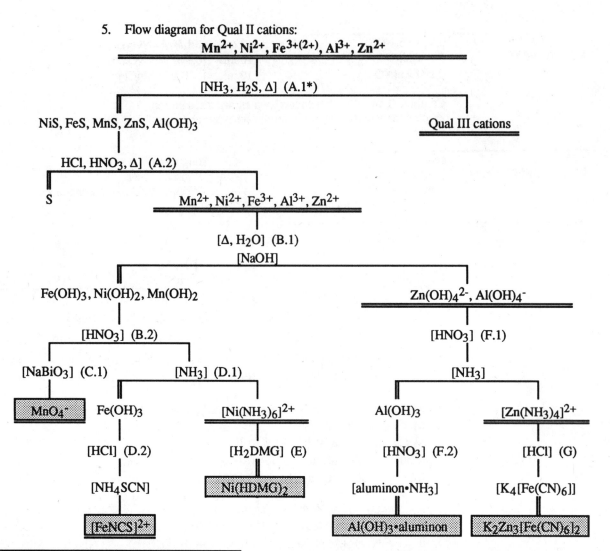

$Mn^{2+}, Ni^{2+}, Fe^{3+(2+)}, Al^{3+}, Zn^{2+}$

[NH₃, H₂S, Δ] (A.1*)

NiS, FeS, MnS, ZnS, Al(OH)₃   Qual III cations

HCl, HNO₃, Δ] (A.2)

S $Mn^{2+}, Ni^{2+}, Fe^{3+}, Al^{3+}, Zn^{2+}$

[Δ, H₂O] (B.1)
[NaOH]

Fe(OH)₃, Ni(OH)₂, Mn(OH)₂ $Zn(OH)_4^{2-}, Al(OH)_4^-$

[HNO₃] (B.2)   [HNO₃] (F.1)

[NaBiO₃] (C.1)   [NH₃] (D.1)   [NH₃]

MnO_4^- Fe(OH)₃ $[Ni(NH_3)_6]^{2+}$ Al(OH)₃ $[Zn(NH_3)_4]^{2+}$

[HCl] (D.2)   [H₂DMG] (E)   [HNO₃] (F.2)   [HCl] (G)

[NH₄SCN]   Ni(HDMG)₂   [aluminon•NH₃]   [K₄[Fe(CN)₆]]

$[FeNCS]^{2+}$ Al(OH)₃•aluminon   K₂Zn₃[Fe(CN)₆]₂

LABORATORY QUESTIONS

1. a. HCl/HNO₃ mixture is used to dissolve the Qual II precipitates.
   b. NaOH separates the Qual II ions into two groups, Mn^{2+}, Ni^{2+} and Fe^{3+} precipitate as the hydroxides, but Zn^{2+} and Al^{3+} form soluble hydroxide polyatomic anions.
   c. NH₃ separates Fe^{3+} and Ni^{2+}; Fe^{3+} precipitates as Fe(OH)₃ in the presence of NH₃, but Ni^{2+} forms the soluble $[Ni(NH_3)_6]^{2+}$ complex.

2. HNO₃ oxidizes the sulfide ion to free sulfur and oxidizes iron(II) ion to iron(III) ion.

*3. Ni^{2+}, Mn^{2+}, Fe^{3+}, and Al^{3+} precipitate with the addition of NH₃; only Zn^{2+} would remain in solution as $[Zn(NH_3)_4]^{2+}$.

4. Both Fe^{3+} and Ni^{2+} precipitate as the hydroxides with the addition of NaOH; a separation of the two ions would not occur.

5. Both Al^{3+} and Zn^{2+} remain in solution as $Al(OH)_4^-$ and $Zn(OH)_4^{2-}$ in the presence of OH⁻; no separation of the two ions would occur.

*6. If no precipitate forms, then Mn^{2+}, Fe^{3+} and Ni^{2+} are absent from the solution. Parts C, D, and E of the Experimental Procedure, the analysis for Mn^{2+}, Fe^{3+} and Ni^{2+}, can be omitted.

*7. Since FeS and NiS are black precipitates, it can be concluded that Fe^{2+} and Ni^{2+} are absent from the test solution.

8. Because the NH_3 forms a blue $[Ni(NH_3)_6]^{2+}$ complex with Ni^{2+}, it can be concluded that Ni^{2+} is absent in the test solution; Fe^{3+} is present from the formation of the red-brown precipitate.

9. a. $S^{2-}(aq)$ in an acidic solution precipitates Cu^{2+}, but not Zn^{2+}.
   b. $HCl(aq)$ will precipitate Ag^+, but not Zn^{2+}.
   c. An excess of NaOH precipitates Mn^{2+}, but not Zn^{2+}.

10. a. Water; $NiCl_2$ dissolves but NiS does not
    b. $Zn(OH)_2$ is amphoteric, an excess NaOH forms $[Zn(OH)_4]^{2-}$, but $Fe(OH)_3$ remains as a precipitate
    c. $HCl(aq)$ dissolves ZnS, but HNO_3, an oxidizing acid, is necessary to dissolve CuS

LABORATORY QUIZ
1. a. Identify a reagent that separates Al^{3+} from Fe^{3+}.   [Answer: excess $OH^-(aq)$]
   b. Identify a reagent that separates Fe^{3+} from Ni^{2+}.   [Answer: excess $NH_3(aq)$]

2. How is the presence of Mn^{2+} confirmed in the Qual II analysis?
   [Answer: oxidation to MnO_4^-]

3. Zinc hydroxide is amphoteric; this means it can react as an acid and as a base. Write an equation that represents each reaction:
   $Zn(OH)_2(s) + H^+(aq) \rightarrow$ [Answer: $Zn^{2+}(aq) + 2 H_2O(l)$]
   $Zn(OH)_2(s) + OH^-(aq) \rightarrow$ [Answer: $Zn(OH)_4^-(aq)$]

4. How are Qual II cations separated from Qual III cations?
   [Answer: H_2S in a neutral to basic solution]

5. How are Zn^{2+} ions separated from Al^{3+} cations?   [Answer: excess $OH^-(aq)$]

Qual III. Mg^{2+}, Ca^{2+}, Ba^{2+}, and General Unknown Examination

INTRODUCTION

After experiencing the analysis procedures for Qual I and Qual II cations, the separation and identification of Qual III cations are welcome. If a General Unknown (a selection of cations from all Qual groups) was assigned, students should have time to complete the Qual III analysis and the General Unknown in a three hour laboratory period.

The chemical principles used in this experiment are ionic equilibrium, precipitation, gaseous evolution, and flame ionization. Ask students to identify these principles as they proceed through the experiment.

WORK ARRANGEMENT

Partners can share observations and discussion on a test solution containing all of the cations; individuals on a test solution containing an unknown number of the Qual III cations and the General Unknown.

TIME REQUIREMENT

1.5 hours for Qual III and 1.5 hours for the General Unknown

LECTURE OUTLINE

1. Follow the Instruction Routine outlined in "To the Laboratory Instructor."

2. Review the principles of the separation and identification of the Qual III cations; the flow diagram for Question 5 on the Prelaboratory Assignment is to be completed before the experiment is begun.

3. If the General Unknown is assigned, do the following:

    Review, assist, and correct each student's personal flow diagram. Procedural details such as drops, stir, decant, centrifuge, color, etc. should *not* be a part of the flow diagram— expect the student to know the *chemistry* of the five cations and the purpose for each procedural step.

    We permit each student the opportunity to check the accuracy in their flow diagram prior to the General Unknown examination. The test solutions, reagents, and equipment for the analyses of the Qual I through Qual III cations are available for this accuracy check. An "open" laboratory time prior to the examination is scheduled.

4. Approve the final flow diagram *before* the student begins the General Unknown analysis. This diagram is used *exclusively* by the student during the testing period; no laboratory manual is permitted.

5. The General Unknown test solution should contain at least two of the five cations that were designated to be in the unknown.

6. One hour is sufficient time for the General Unknown analysis. There should be *no* discussions during this time period—this is an examination!

7. If this is the last laboratory period for the course, proceed through a formal check-out procedure. Advise students of the procedure.

CAUTIONS & DISPOSAL

*   Acids and Bases. $6\,M$ and conc HCl, $6\,M$ H_2SO_4, and $6\,M$ NH_3 are used in this experiment. For a General Unknown, a larger array of acids and bases will be available for student use.
    Be prepared to clean up acid and base spills; have sodium bicarbonate available in the laboratory.

- Students should dispose of their test solutions in a "Waste Metal Salts" container.

TEACHING HINTS

1. **Part A.** If the unknown sample only contains Qual III cations, Part A can be omitted.

2. **Part B.1.** Students often confuse the yellow the solution as being confirmatory for Ba^{2+}.

3. **Part B.3.** A second confirmatory test for Ba^{2+} is optional.

4. **Part D.1.** The precipitate for the confirmatory test for Mg^{2+} may be slow in forming. Be patient.

5. Data of the tests for the various steps of the Experimental Procedure.

Procedure Number and Ion	Test Reagent or Technique	Observation (Color or General Appearance)		Chemical(s) Responsible for Observation	Equation(s) for Observed Reaction
#1 Ba^{2+}	K_2CrO_4	ppt	yellow	$BaCrO_4$	36.1
#2	flame		green/ yellow	—	—
#3	H_2SO_4	ppt	white	$BaSO_4$	36.2
#4 Ca^{2+}	$NH_3/K_2C_2O_4$	ppt	white	CaC_2O_4	36.3
#5	flame		red/ orange	—	—
#6 Mg^{2+}	$NH_3/$ Na_2HPO_4	ppt	white	$MgNH_4PO_4$	36.4

CHEMICALS REQUIRED

6 M HCl (dropper bottle)	2 mL	6 M NH_3 (dropper bottle)	1 mL
1 M K_2CrO_4 (dropper bottle)	1 mL	1 M $K_2C_2O_4$ (dropper bottle)	1 mL
conc HCl (dropper bottle)	1 mL	1 M Na_2HPO_4 (dropper bottle)	0.5 mL
6 M H_2SO_4 (dropper bottle)	1 mL		

GENERAL UNKNOWN CHEMICALS

All chemicals that were required for the Qual I, II, and III analyses need to be available for the General Unknown Examination.

SUGGESTED UNKNOWNS

0.1 M $Ba(NO_3)_2$	10 mL
0.3 M $Ca(NO_3)_2$	10 mL
0.1 M $Mg(NO_3)_2$	10 mL

Combine 2 mL of each solution for a *known* test solution.
Prepare the *unknown* test solution by substituting 2 mL of deionized water for each anion *not* present.

SPECIAL EQUIPMENT

coiled flame test wire sealed in glass tubing	Bunsen burner
centrifuge	wire gauze
evaporating dish	ring stand and iron support rings 2
	"Waste Metal Salts" container

PRELABORATORY ASSIGNMENT

1.

Equation	Mass Action Expression	K_s Value	Color
$BaCrO_4(s) \rightleftharpoons Ba^{2+}(aq) + CrO_4^{2-}(aq)$	$K_s = [Ba^{2+}][CrO_4^{2-}]$	2.4×10^{-10}	yellow
$BaSO_4(s) \rightleftharpoons Ba^{2+}(aq) + SO_4^{2-}(aq)$	$K_s = [Ba^{2+}][SO_4^{2-}]$	1.5×10^{-9}	white
$CaC_2O_4(s) \rightleftharpoons Ca^{2+}(aq) + C_2O_4^{2-}(aq)$	$K_s = [Ca^{2+}][C_2O_4^{2-}]$	1.8×10^{-9}	white

Molar solubility of $BaCrO_4 = 1.5 \times 10^{-5}$ mol/L, least soluble
Molar solubility of $BaSO_4 = 3.9 \times 10^{-5}$ mol/L
Molar solubility of $CaC_2O_4 = 4.2 \times 10^{-5}$ mol/L, most soluble

2. a. $K_2CrO_4(aq)$ precipitates Ba^{2+}, but not Ca^{2+}
    b. $H_2S(aq)$ or thioacetamide with heat precipitates Fe^{2+}, but not Ba^{2+}
    c. $K_2C_2O_4(aq)$ precipitates Ca^{2+}, but not Mg^{2+}

3. a. $K_2CrO_4(aq)$ precipitates Ba^{2+}, but not Ca^{2+}
    b. $H_2S(aq)$ or thioacetamide with heat precipitates Ni^{2+}, but not Ba^{2+}
    c. $K_2C_2O_4$, NaOH, . . .precipitates Ca^{2+}, but not Na^+

4. a. Ca^{2+}
    b. Ba^{2+}
    c. Mg^{2+}

5. Flow diagram for the Qual III cations.

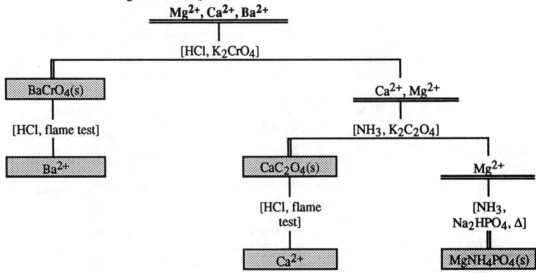

LABORATORY QUESTIONS

1. $BaCl_2$ is soluble, whereas $BaSO_4$ is insoluble. If HCl were used instead of H_2SO_4, no precipitate forms.

2. BaC_2O_4 is soluble, whereas $BaCrO_4$ is insoluble. If $K_2C_2O_4$ is used, Ba^{2+} does *not* precipitate, but CaC_2O_4 does precipitate.

3. One can only conclude that Ba^{2+} is absent from the test solution; a yellow solution is not a confirmatory test.

4. Upon addition of NaOH, the formation of insoluble $Mg(OH)_2$ is immediate.

5. Flow diagram for Ag^+, Cu^{2+}, and Mn^{2+}.

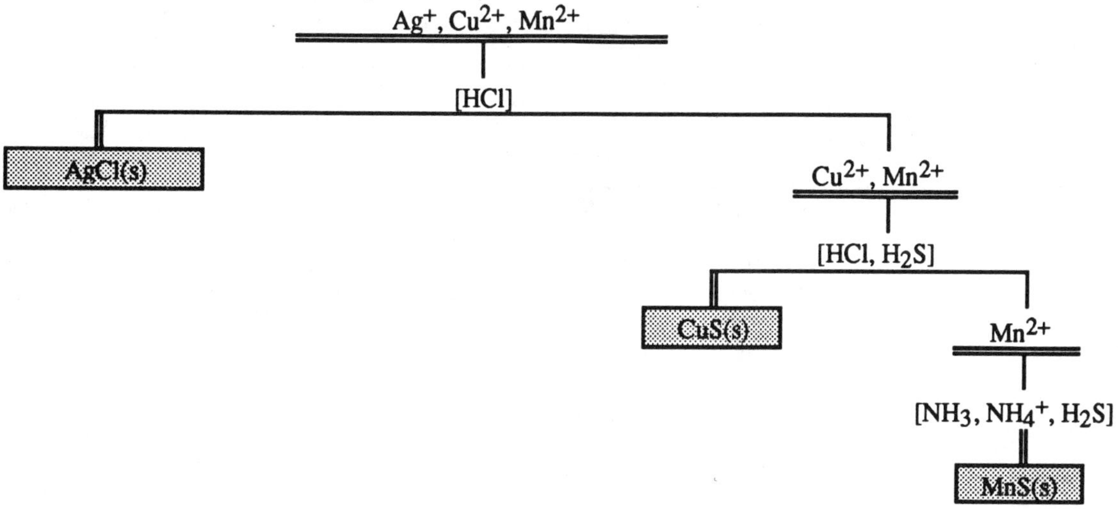

Ag^+, Cu^{2+}, Mn^{2+}

[HCl]

AgCl(s)

Cu^{2+}, Mn^{2+}

[HCl, H_2S]

CuS(s)

Mn^{2+}

[NH_3, NH_4^+, H_2S]

MnS(s)

LABORATORY QUIZ

1. Identify the Qual I and Qual III cations that can be identified with a flame test.

[Answer: Na^+, K^+, Ba^{2+}, Ca^{2+}]

2. What reagent separates Ba^{2+} cations from Ca^{2+} cations?   [Answer: CrO_4^{2-}]

3. List the reagent(s) used to separate:
   a. Qual I cations from Qual II and III cations.   [Answer: $H_2S(aq)$, acidic solution]
   b. Qual II cations from Qual III cations. [Answer: $H_2S(aq)$, basic to neutral solution]

4. List the cations, as studied in this course, that are
   a. Qual I cations   [Answer: Na^+, K^+, NH_4^+, Ag^+, Cu^{2+}, Bi^{3+}]
   b. Qual II cations   [Answer: Mn^{2+}, Ni^{2+}, $Fe^{3+}(Fe^{2+})$, Al^{3+}, Zn^{2+}]
   c. Qual III cations   [Answer: Mg^{2+}, Ca^{2+}, Ba^{2+}]

5. Identify a reagent that separates
   a. Cu^{2+} from Ba^{2+} [Answer: $H_2S(aq)$, acidic solution]
   b. Ag^+ from Cu^{2+} [Answer: $HCl(aq)$]
   c. Fe^{3+} from Ni^{2+} [Answer: excess $NH_3(aq)$]

Synthesis of an Alum

INTRODUCTION

In our general chemistry program, we require each of our students to synthesize at least one compound during the year. There are three synthesis experiments in this manual: an aspirin synthesis (Experiment 29), three alum syntheses (Experiment 37), the synthesis of three coordination compounds (Experiment 38), and the extensive synthesis and analysis of the *tris*oxalatoferrate(III) complex (Experiment 39A,B,C). Whenever possible, we have the student select one of personal interest; this encourages an exchange of observations and results with students performing other preps.

In your assignment of a laboratory synthesis for the course, do *not* allow a student to perform a preparation for which there is not proper safety equipment. Also do not assign a difficult preparation to a student who has limited laboratory skills: the alums are easiest (specifically the potassium alum is easiest) to prepare, followed by aspirin, and then the coordination compounds. Experiment 39 requires 9–15 hours of laboratory time to complete the experiment in its entirety.

Depending upon the time available, we request *at most* the preparation of two alums in one laboratory period. The alums are relatively simple to prepare with good yields; the ferric alum is the most difficult to synthesize.

WORK ARRANGEMENT

Individuals.

TIME REQUIREMENT

2 hours per alum; however, because of the overnight "waiting time" for full crystallization to occur, two preparations can easily be completed during a 3-hour laboratory period.

LECTURE OUTLINE

1. Follow the Instruction Routine outlined in "To the Laboratory Instructor."

2. Indicate which alums are to be synthesized.

3. Showing a calculation for a theoretical yield of an alum is time-saving.

CAUTIONS & DISPOSAL

- **Part A.** 4 M KOH reacts with the skin to form a soap. Avoid skin contact. H_2 gas is evolved in the reaction of aluminum with KOH. Conduct the experiment in a well-ventilated area. Caution is advised in the handling of the 6 M H_2SO_4.

- **Part B.** Handle the 2 M H_2SO_4 and the $K_2Cr_2O_7$ with care. The addition of ethanol to the dichromate solution (Part B.2) should be done in a fume hood. Potassium dichromate in sulfuric acid is a *strong* oxidant!

- **Part C.** Avoid skin contact with the 2 M H_2SO_4 and conc HNO_3. The addition of conc HNO_3 (Part C.2) is exothermic and rapid. NO_2 fumes are evolved in the reaction; therefore, the reaction should be completed in a fume hood.

- Dispose of all filtrates in a "Waste Salts" container.

TEACHING HINTS

1. **Part A.2.** The reaction of 1 g of aluminum metal with 4 M KOH takes about 20 minutes. Students should constantly swirl the contents.
   High porosity filter paper should be used when filtering reaction mixture. Glass wool may be used as a filter.

2. **Part A.3.** The reaction of 6 M H_2SO_4 with 4 M KOH is very exothermic; therefore, the acid should be added *slowly with stirring*. Judgment is to be made in adding the amount of H_2SO_4—add until the $Al(OH)_3$ forms, but not to the point where the $Al(OH)_3$ dissolves. If too much H_2SO_4 is accidentally added, merely add KOH.

3. **Part A.5.** The white crystals form upon cooling; an ice bath accelerates the crystallization. If crystallization does not occur upon cooling, reduce the volume of the liquid by one-half with a low heat.

4. **Part B.2.** The reaction of the $Cr_2O_7^{2-}$ with ethanol is very exothermic. "*Cautiously* and *slowly*, add *drops* of 95% ethanol with constant stirring . . . " is to be taken seriously!

5. **Parts B.2, 3.** Crystallization does not always readily occur. If not, again reduce the volume of the liquid over low heat. The black-green crystals are difficult to see because of the black-green solution—patience is a virtue in this crystallization.

6. **Part C.2.** The addition of conc HNO_3 should be done carefully.

7. **Part C.3.** The off-white crystals of the ferric alum are tinted with a rust color from the ferric ion.

8. **Part D.2.** Inspect the apparatus for the melting point measurement (Figures 2.2, 2.3 and 37.3). An iron ring placed around the water bath will decrease the possibility of the beaker of water from being accidentally knocked off the wire gauze.

9. **Part D.3.** For the alums in this experiment:

Alum	Molar Mass (g/mol)	Melting Point
$KAl(SO_4)_2 \cdot 12H_2O$	474.39	92.5°C
$KCr(SO_4)_2 \cdot 12H_2O$	499.41	89°C
$NH_4Fe(SO_4)_2 \cdot 12H_2O$	482.19	39-41°C

CHEMICALS REQUIRED

All amounts are on a "per trial" basis.

Potassium alum		Chrome alum		Ferric alum	
aluminum (beverage can/foil)	1 g	2.0 M H_2SO_4	100 mL	2.0 M H_2SO_4	15 mL
4 M KOH	25 mL	$K_2Cr_2O_7(s)$	15 g	$FeSO_4 \cdot 7H_2O(s)$	10 g
6 M H_2SO_4	30 mL	95% ethanol	10 mL	conc HNO_3	5 mL
50% ethanol	10 mL	50% ethanol	10 mL	$(NH_4)_2SO_4(s)$	5 g
				50% ethanol	10 mL

SPECIAL EQUIPMENT

balance (±0.01 g)		watchglass	
fume hood		Bunsen burner	
Büchner funnel and flask	1	thermometer clamp	
melting point capillary tube	1/trial	ring stand and iron support rings	2
hot plate	1/10 students	rubber band	
thermometer, 110°C	1	"Waste Salts" container	
ice		"Waste Mercury" container	

PRELABORATORY ASSIGNMENT

1. An alum is a hydrated double sulfate salt with a general formula of $M^+M^{3+}(SO_4)_2 \cdot 12H_2O$.

2. $Al(OH)_3$ is amphoteric because it reacts with an acid,
   $Al(OH)_3(s) + 3 H_3O^+(l) \rightleftharpoons Al(H_2O)_6^{3+}(aq)$
   or it reacts with a base,
   $Al(OH)_3(aq) + OH^-(aq) \rightleftharpoons Al(OH)_4^-(aq)$

3. a. *Yes.* The 12 moles of water molecules per mole of double salt are a part of the crystalline structure of the alum and therefore are considered as a part of the molar mass of the alum and (product) are to be used in calculating the theoretical yield in the reaction.
   b. $KAl(SO_4)_2 \cdot 12H_2O$, 474.32 g/mol

4. 18.6 g $KAl(SO_4)_2 \cdot 12H_2O$

5. 52.62 g $KCr(SO_4)_2 \cdot 12H_2O$

6. $Al_2(SO_4)_3 \cdot 18H_2O$ is the limiting reactant producing 37.03 g $NH_4Al(SO_4)_2 \cdot 12H_2O$

7. Heating the water bath too rapidly may cause the melting point to be surpassed before it is recorded. As a result, the melting point may be recorded *too high*.

LABORATORY QUESTIONS

1. The greater the surface area of the aluminum pieces, the more rapidly is its reaction with the KOH, a kinetics factor that relates to the state of subdivision of the reactants.

2. a. $Al(OH)_3$
   b. $Al(OH)_3$ dissolves in Part A.4 because its solubility increases with an increase in temperature.

*3. Additional time at a low temperature allows for the solid alum to establish an equilibrium with its ions in solution and form larger and more "perfect" crystals.

4. See Experiment 6. Heat the alum crystals to a high temperature in a crucible—the crystals change their appearance because of the loss of the hydrated water molecules.

5. The presence of the impurities lowers the melting point of a pure substance—this observation is a result of the colligative property of the solute.

LABORATORY QUIZ

1. a. Write the general formula of an alum. [Answer: $M^+M^{3+}(SO_4)_2 \cdot 12H_2O$]
   b. Write the formula of the alum you prepared in the laboratory.

2. What is the theoretical yield of $NH_4Fe(SO_4)_2 \cdot 12H_2O$ (molar mass = 482.19 g/mol) from the reaction of 0.79 g of iron and an excess of all other reactants? [Answer: 6.8 g]

3. What procedure can be used if crystallization of the alum does not occur, even in an ice bath? [Answer: reduce the volume of the solution]

4. The ferric alum is prepared with $FeSO_4 \cdot 7H_2O$ as a reactant. What substance is used to oxidize the iron(II) ion to iron(III) ion? [Answer: $HNO_3(aq)$]

5. Three moles of ethanol (C_2H_5OH, density is 0.789 g/mL) reduce one mole of potassium dichromate, $K_2Cr_2O_7$, in preparing chrome alum. Determine the volume, in mL, of ethanol required to react with 15.0 g of $K_2Cr_2O_7$.
   $8 H^+(aq) + Cr_2O_7{}^{2-}(aq) + 3 C_2H_5OH(aq) \rightarrow$
   $3 CH_3CHO(aq) + 2 Cr^{3+}(aq) + 7 H_2O(l)$ [Answer: 8.94 mL]

Transition Metal Chemistry

INTRODUCTION	A qualitative study of the coordination chemistry of the Cu^{2+}, Ni^{2+}, and Co^{2+} ions and the synthesis and isolation of several Cu^{2+} and Ni^{2+} coordination compounds are presented in this experiment. Various ligands are added to aqueous solutions containing the metal ions to observe color changes and to explain the stability of the resulting complexes.
	While this experiment is mostly qualitative, students enjoy the various colors for the complexes and discuss, at length, the relative ligand strengths. Most of all, students should be patient with the tests in this experiment; oftentimes students need to make a decision on an observation and/or on the amount of reagent to use.
WORK ARRANGEMENT	Partners. A discussion of the procedure and an observation provides a more valuable learning experience in this laboratory experiment.
TIME REQUIREMENT	Parts A–D require 1.5 hours, Parts E–G require 2.5 hours. Generally, for a 3 hour laboratory period, we permit the student to select one synthesis (Part E, F, or G) of a coordination compound.

LECTURE OUTLINE

1. Follow the Instruction Routine outlined in "To the Laboratory Instructor."

2. Discuss or define the terms complex, ligand, coordination number, chelating agent—these terms are all relatively new terms for students. Use a number of examples in your discussion.

3. **Parts A–D.** Review the introduction to the Experimental Procedure—students may need to add more ligand, withdraw some of the metal ion solution from a well, add an excess of ligand, etc. to obtain a color change for an observable complex formation.

4. A 24-well tray is suggested for use in this experiment—the volumes of solutions are small and the test comparisons are easy. If a 24-well tray is *not* available, from 5 to 7 75-mm test tubes should be available.

CAUTIONS & DISPOSAL

- Acids and Bases. Conc HCl , conc NH_3, and ethylenediamine are used in this experiment.
  Be prepared to effectively clean up acid/base spills. A supply of sodium bicarbonate should be present in the laboratory.

- The test solutions should be disposed in a "Waste Metal Ions Solutions" container.

TEACHING HINTS

1. Students need to be encouraged to work with these solutions—discover by doing. A bit of research philosophy should be encouraged for this experiment.

2. While students are not required to write equations in this experiment, you may wish to write a few on the chalkboard to illustrate what is happening in these reactions.

3. A summary of the observations from Parts A–D follow.

A. Chloro Complexes of the Copper(II), Nickel(II), and the Cobalt(II) Ions

Solution	Color/H_2O	Color/HCl	Formula of Complex	Effect of H_2O
0.1 M $CuSO_4$	blue	yellow	$[CuCl_4]^{2-}$	blue
0.1 M $Ni(NO_3)_2$	green	lt. yellow	$[NiCl_6]^{4-}$	lt. green
0.1 M $CoCl_2$	red (dark pink)	blue	$[CoCl_6]^{4-}$ (actually $[CoCl_4]^{2-}$)	pink

2. In each case state whether the aqua complex or the chloro complex is more stable: Cu^{2+}aqua; Ni^{2+}aqua; Co^{2+}aqua

B. The Complexes of the Copper(II) Ion

Ligand	Color	Formula of Complex	Effect of OH⁻
NH_3	dark blue	$[Cu(NH_3)_4]^{2+}$	none
ethylenediamine	dark lavender	$[Cu(en)_2]^{2+}$	none
SCN^-	yellow/green	$[Cu(NCS)_4]^{2-}$	ppt
H_2O	blue	$[Cu(H_2O)_4]^{2+}$	ppt

C. The Complexes of the Nickel(II) Ion

Ligand	Color	Formula of Complex	Effect of OH⁻
NH_3	light blue	$[Ni(NH_3)_6]^{2+}$	ppt
ethylenediamine	lt. lavender	$[Ni(en)_3]^{2+}$	none
SCN^-	no change	$[Ni(NCS)_6]^{4-}$	ppt
H_2O	lt. green	$[Ni(H_2O)_6]^{2+}$	ppt

D. The Complexes of the Cobalt(II) Ion

Ligand	Color	Formula of Complex	Effect of OH⁻
NH_3	amber	$[Co(NH_3)_6]^{2+}$	ppt
ethylenediamine	slow to develop amber	$[Co(en)_3]^{2+}$	none
SCN^-	intensifies red, no change	$[Co(NCS)_6]^{4-}$	ppt
H_2O	dark pink	$[Co(H_2O)_6]^{2+}$	ppt

Summary of Data from Parts A–D:
The complexes that showed a precipitate with the addition of 1 M NaOH:

Ligand/Metal	Cu^{2+}/OH⁻	Ni^{2+}/OH⁻	Co^{2+}/OH⁻
NH_3		√	√
ethylenediamine			
SCN^-	√	√	√
H_2O	√	√	√

4. **Part E.** This synthesis is relatively simple and can be completed by nearly all students regardless of laboratory experience.

5. **Part E.2 (also Parts F.2 and G.2).** The cool 95% ethanol washes the precipitate with a minimum dissolving of the product. The coordination compound is more soluble in room temperature ethanol and more soluble in water.

6. **Part F.** This synthesis, too, is relatively simple. The product is light lavender.

7. **Part G.** This synthesis requires only one step and the product is a deep lavender.

8. You should approve the synthesis of the coordination compound that is synthesized by the student.

CHEMICALS REQUIRED	Parts A–D		conc NH₃ (dropper bottle)	2 mL
	0.1 M CuSO₄ or Cu(NO₃)₂	3 mL	ethylenediamine	2 mL
	0.1 M NiCl₂ or Ni(NO₃)₂	3 mL	0.1 M KSCN	2 mL
	0.1 M CoCl₂ or Co(NO₃)₂	3 mL	1 M NaOH	4 mL
	conc HCl	2 mL		

Parts F–G

	[Cu(NH₃)₄]SO₄	[Ni(NH₃)₆]Cl₂	[Ni(en)₃]Cl₂·2H₂O
	$CuSO_4\cdot5H_2O(s)$ 6 g	$NiCl_2\cdot6H_2O(s)$ 6 g	$NiCl_2\cdot6H_2O(s)$ 6 g
	conc NH₃ 15 mL	conc NH₃ 20 mL	ethylenediamine 5 g
	95% ethanol 30 mL	95% ethanol 25 mL	95% ethanol 25 mL
	acetone 5 mL	acetone 5 mL	acetone 5 mL

SPECIAL EQUIPMENT

24-well plate and Beral pipets		balance (±0.01 g)
ice		ring stand and iron support rings 2
Büchner funnel and flask	1	flask clamp
Bunsen burner		"Waste Metal Ion Solutions" container
watchglasses	1–2	

PRELABORATORY ASSIGNMENT

1. A complex consists of a groups of molecules and/or anions, called ligands, bonded to a metal ion.

2. a. A ligand is the molecule and/or anion, which is a Lewis base, bonded to the metal ion in a complex.
   b. A monodentate ligand forms one bond to the metal ion and a bidentate ligand forms two bonds to the metal ion in the formation of a complex.

*3. Ligands alter the electronic energy levels in the metal ion; a change of electronic energy levels changes the energy (and wavelength) of light that the metal ion absorbs and, therefore, the energy (and wavelength) of light that is transmitted.

4. A chromium ion that has a coordination number of 6 in a complex has 6 bonds between the Cr^{3+} ion and its ligands.

5. a. $[Cu(H_2O)_4]^{2+}$
   b. $[Ni(en)_3]^{2+}$
   c. $[FeF_6]^{3-}$

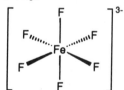

6. a. ammonia and chloride ion
   b. ammonia
   c. ammonia and water

LABORATORY QUESTIONS

1. a. The aqua complex of Cu^{2+} is more stable than the chloro complex.
   b. The ethylenediamine complex of Ni^{2+} is more stable (no precipitate with OH^-) because of its chelating effect.
   c. The ethylenediamine complex ion of Co^{2+} is more stable (no precipitate with OH^-) because of its chelating effect.

2. a. $H_2NCH_2CH_2NH_2$
   b. Cl^-

3. a. $Cr(OH)_3$
   b. $[Cr(OH)_4]^-$

4. a. $[Fe(NCS)_6]^{3-}$, $[FeF_6]^{3-}$
   b. $[FeF_6]^{3-}$ is the more stable complex.

5. 95% ethanol is less polar than is water. Since the coordination compound is ionic, its solubility will be less in a less polar solvent. The washing of the compound with a less polar solvent will result in smaller amounts of product being lost in the wash cycle.

1. a. Identify the ligands in the complex, $[Cr(NH_3)_4(OH)_2]^+$.   [Answer: NH_3, OH^-]
    b. What is the oxidation number of chromium in this complex?   [Answer: 3^+]
    c. What is the coordination number of chromium in this complex?   [Answer: 6]
    d. Sketch the three-dimensional shape of the complex. Sketch all the isomers of the complex.

2. Which complex is predicted to be more stable, $[Co(en)_6]^{2+}$ or $[Co(NH_3)_6]^{2+}$? Explain.
    [Answer: $[Co(en)_3]^{2+}$ because (en) is a bidentate]

3. Indicate the color of the aqua complexes of
    a. Ni^{2+} d. Cu^{2+}
    b. Co^{2+} e. Fe^{3+}
    c. Zn^{2+} [Answer: green, pink, colorless, sky-blue, orange, respectively]

4. Distinguish between a monodentate ligand and a bidentate ligand.
    [Answer: a monodentate ligand bonds once to the metal ion, a bidentate ligand bonds twice]

5. Conc NH_3 is added to an aqueous solution of $[Ni(H_2O)_6]Cl_2$ to prepare $[Ni(NH_3)_6]Cl_2$. Which is the stronger ligand, NH_3 or H_2O?   [Answer: NH_3]

6. Ligands A and B form the complexes $[MA_6]^{2+}$ and $[MB_6]^{2+}$, where M is a divalent metal ion. When NaOH is added to separate solutions containing the complexes, a precipitate forms in the solution containing $[MB_6]^{2+}$ complex, but not in the solution containing the $[MA_6]^{2+}$ complex.
    a. Write the formula for the precipitate.   [Answer: $M(OH)_2$]
    b. Which ligand is the stronger ligand?   [Answer: A]
    c. Which complex is more stable?   [Answer: $[MA_6]^{2+}$]

Experiment 39

39A: Synthesis of K$_w$Fe$_x$(C$_2$O$_4$)$_y$(H$_2$O)$_z$, a Coordination Compound
39B: Analysis for Oxalate in K$_w$Fe$_x$(C$_2$O$_4$)$_y$(H$_2$O)$_z$
39C: Analysis for Iron(III) in K$_w$Fe$_x$(C$_2$O$_4$)$_y$(H$_2$O)$_z$

INTRODUCTION

Experiment 39 is "project" experiment, requiring *at least* three laboratory periods for the synthesis and *complete* analysis of the coordination compound. The experiment incorporates several techniques and revisits many of the principles used in earlier experiments. For these reasons, "39" is an excellent wrap-up experiment for the general chemistry laboratory program.

There is flexibility in addressing this group of three experiments:

Experiment 39A can stand alone and be used for the mere synthesis of a coordination compound; the synthesis is nearly foolproof. Experiments 39B and/or 39C can be completed separately for the analysis of the coordination compound. If values of w, x, y, and z are to be assigned for the determination of the empirical formula and both 39B and 39C are not completed, then explicit information about the values of x (Expt 39C) and/or y (Expt 39B) must be given to students.

WORK ARRANGEMENT

The work arrangement may vary during the experiment. We require each student to synthesize the coordination compound and then to cooperate with others during the analysis of the oxalate and iron(III) ions.

TIME REQUIREMENT

6–8 hours for the synthesis and analysis, although the time required for the overnight crystallization, setup time for the chemical analyses and the data analysis varies with each student. Experiment 39A requires about 2 hours (through Part B.4); Experiment 39B requires about 3 hours; Experiment 39C requires about 2.5 hours. We allow three laboratory periods for the synthesis and analyses and outside time for the calculations.

LECTURE OUTLINE

1. Follow the Instruction Routine outlined in "To the Laboratory Instructor."

2. Since Experiment 39A does not require a full 3-hour laboratory period, there is time at the beginning of the first period to present an overview of the "project" and the somewhat open-ended experiment.
   - Experiment 39A outlines the procedure for the synthesis of K$_w$Fe$_x$(C$_2$O$_4$)$_y$(H$_2$O)$_z$.
   - Experiment 39B describes the titrimetric procedure for the analysis of the oxalate ion using KMnO$_4$ as the titrant.
   - Experiment 39C describes the spectrophotometric analysis of the iron(III) ion, reduced to iron(II) ion with ascorbic acid, as the *tris*-bipyridyliron(II) complex.

3. Explain that the values of w, x, y, and z may vary with the synthesis; students should *not* believe that there is a "correct" formula for the coordination compound.

4. The three Report Sheets for this experiment are self-designed. Advise students that their data must be organized and easy to interpret. Careful, meticulous data collection and recording are absolutely necessary.

5. Emphasize that the data collection is cumulative: data collected in Experiment 39A is used and evaluated at the conclusion of Experiment 39C; therefore, organized and explicit data must be consistently recorded for interpretation.

6. It is best *not* to discuss all of Experiment 39 in one session; rather use a portion of each laboratory period that extends over the several sessions to discuss the experiment.

<table>
<tr><td>**CAUTIONS & DISPOSAL**</td><td>

Experiment 39A

• **Part A.** 6 M H_2SO_4 should be handled carefully. Acetone (flammable) and 6% H_2O_2 are to be handled with caution.

• Oxalic acid is toxic. Students should wash their hands if there is any skin contact.

• The supernatant from Part A and the filtrate from Part B can be discarded in the "Waste Acids" container.

Experiment 39B

• Part 3. Caution students on the handling of 6 M H_2SO_4.

• All test solutions are to be discarded in the "Waste Salts" container and the excess $KMnO_4$ in the buret is to be discarded in the "Waste Oxidants" container after the analysis is complete.

Experiment 39C

• Provide a "Waste Bipyridyliron(II)" container

• 6 M H_2SO_4 is used in Part B.1.

</td></tr>
<tr><td>**TEACHING HINTS**</td><td>

Experiment 39A

1. **Parts B.1 and B.2.** Two chemical processes should be considered: (1) enough peroxide must be added to oxidize all of the Fe^{2+} to Fe^{3+} and (2) additional oxalate ion must be added for complexing to the Fe^{3+} ion. If the procedure is closely followed, the complex should be $[Fe(C_2O_4)_3]^{3-}$. The appearances of the yellow-green solution (Part B.2) and the emerald green crystals (Part B.3) are nearly foolproof.

2. **Part B.3.** The ethanol reduces the polarity of the mixed solvent system and reduces the solubility of the coordination compound.

3. **Part B.3.** Note that the coordination compound is light sensitive.

4. **Part B.4.** The yield of the coordination compound is most often great enough to allow for the analyses in Experiments 39B and 39C, with additional crystals left over for "show."

Experiment 39B

1. The standardized $KMnO_4$ solution should be protected from the light.

2. A total of about 0.3 g of the coordination compound are required for the oxalate analyses (Experiment 39B). An additional 0.1 g is necessary for the iron(III) analysis in Experiment 39C. Therefore if the mass of crystals is marginal for the analyses, reduce the mass of compound for each trial in the oxalate analysis (Experiment 39B).

3. **Part 4.** The reaction of MnO_4^- with $C_2O_4^{2-}$ is slow; the application of heat ($\approx 60°C$) increases the reaction rate and removes the CO_2 as it is formed. In addition the reaction is catalyzed by Mn^{2+}, a reduction product of MnO_4^-; as Mn^{2+} is not initially present in the reaction system the initial reaction rate is slow.

4. **Part 4.** At this stage of laboratory maturity, students should be able to add half-drops of titrant.

</td></tr>
</table>

5. At the completion of the analysis, advise students to thoroughly clean the buret before it is returned or stored.

Experiment 39C

1. **Part A.** An absorbance versus $[Fe(bipy)_3]^{2+}$ plot for a set of standard solutions is prepared. The plot is used in Part B to determine the amount of iron in a measured sample of the coordination compound. The λ_{max} for $[Fe(bipy)_3]^{2+}$ is 520 nm.

2. Note the calculation that is required for Part A.3.

3. Explain and oversee the operation of the spectrophotometers. Advise students to place only "dry" cuvets in the cuvet holder. The %T readings, rather than absorbance readings, should be recorded from the spectrophotometer for more exact interpolations.

4. **Part A.8.** The best *straight* line is to be drawn through the data points.

5. **Part B.1.** A voluminous amount of precipitate forms with the addition of the $CaCl_2$ solution to the solution of the coordination compound—both CaC_2O_4 and $CaSO_4$ (SO_4^{2-} from the sulfuric acid) precipitate.

6. **Part B.4.** At least three of the test solutions from Part B must have absorbance readings that lie within the range of the standard solutions. Students will need to analyze their data and determine a procedure for possibly preparing additional $[Fe(bipy)_3]^{2+}$ solutions for analysis.

7. **Part B.5.** Students may require some assistance in determining the molar concentrations of $[Fe(bipy)_3]^{2+}$ of their test solutions using the plot of the standard solutions (Part A).

8. **Parts B.6 and B.7.** Students will undoubtedly require some assistance with the calculations.

Experiment 39, A Summary of Data

1. Many students will undoubtedly require some assistance in the calculations for w, x, y and z.

2. The basis for the calculation is a 100 g sample. The purpose of the 100 grams is that a reference mass must be established in comparing data from different experiments.

3. Your review of Experiment 39B, Lab Questions 6 will help in formalizing your instruction and assistance. A second challenge is to solve Laboratory Quiz, Question 5. These two problems will give you the experience to "coach" your students through the calculations.

CHEMICALS REQUIRED	**Experiment 39A**		**Experiment 39B**	
	$Fe(NH_4)_2(SO_4)_2 \cdot 6H_2O(s)$	1 g	0.01 M KMnO$_4$ (standardized)	75 mL
	6 M H$_2$SO$_4$	2 mL	6 M H$_2$SO$_4$	15 mL
	1 M H$_2$C$_2$O$_4$	7 mL		
	2 M K$_2$C$_2$O$_4$	4 mL	**Experiment 39C**	
	6% H$_2$O$_2$	9 mL	$Fe(NH_4)_2(SO_4)_2 \cdot 6H_2O(s)$	0.12 g
	95% ethanol	7 mL	ascorbic acid(s)	0.2 g
	75/25 alcohol/water	5 mL	0.0020 M bipyridyl	40 mL
	acetone	10 mL	buffer solution (pH ≈ 4.7)	75 mL
			6 M H$_2$SO$_4$	5 mL
			0.5 M CaCl$_2$	20 mL

SPECIAL EQUIPMENT	**Experiment 39A**		Bunsen burner
	balance (±0.001 g)		"Waste Oxidants" container
	weighing paper	2–3	"Waste Salts" container
	aluminum foil		
	ring stand and irons rings	2	**Experiment 39C**
	200-mm test tube		balance (±0.001 g)
	Bunsen burner		weighing paper 2–3
	Büchner funnel and flask	1	100-mL volumetric flask 2
	"Waste Acids" container		50-mL volumetric flask 2
			2-mL pipet and pipet bulb 1
	Experiment 39B		10-mL graduated pipet 2
	balance (±0.001 g)		spectrophotometer and ≥2 cuvets
	weighing paper	2–3	lint free tissue (Kimwipes)
	aluminum foil		ring stand and iron rings 2
	50-mL buret and buret brush	1	10-mL pipet 1
	ring stand and buret clamp		"Waste *tris*-Bipyridyl" container

PRELABORATORY ASSIGNMENT, EXPERIMENT 39A

1. a. The complex is $[CrCl_2(H_2O)_4]^+$; the charge on the complex ion is 1^+.
   b. The oxidation number of chromium is 3^+ and the coordination number of chromium is 6.
   c. The ligands in the complex are Cl^- and H_2O.

2. The oxidation number of rhodium in $[RhCl(OH)(C_2O_4)_2]^-$ is 5^+ and the coordination number is 6. The oxalate ion is a bidentate.

3. a. Molar mass of $Fe(NH_4)_2(SO_4)_2 \cdot 6H_2O$ is 392.15 g/mol. For 1.0 g sample there is 2.6×10^{-3} mol iron(II) ion.
   b. 5.0×10^{-3} mol $C_2O_4^{2-}$ ion
   c. The limiting reactant is the iron(II) ion.

4. a. Hydrogen peroxide oxidizes Fe^{2+} to Fe^{3+}.
   b. $H_2O_2(aq) + 2H^+(aq) + 2e^- \rightarrow 2H_2O(l)$

5. Ethanol is less polar than is water. Because ionic salts are *very* polar, they become less soluble is less polar solvents.

6. The emerald green crystals of the coordination compound are light sensitive; therefore, the crystals should be protected from the light.

PRELABORATORY ASSIGNMENT, EXPERIMENT 39B

1. Dissolve 0.158 g $KMnO_4$ in a 100-mL volumetric flask and dilute to volume.

2. $0.0662 \text{ g Na}_2\text{C}_2\text{O}_4 \times \dfrac{\text{mol Na}_2\text{C}_2\text{O}_4}{134.02 \text{ g Na}_2\text{C}_2\text{O}_4} \times \dfrac{2 \text{ mol KMnO}_4}{5 \text{ mol Na}_2\text{C}_2\text{O}_4}$

   $= 1.98 \times 10^{-4} \text{ mol KMnO}_4$

   $\dfrac{1.98 \times 10^{-4} \text{ mol KMnO}_4}{0.0183 \text{ L}} = 0.0108 \text{ M KMnO}_4$

3. a. 2.37×10^{-4} mol $C_2O_4^{2-}$ and 0.0209 g $C_2O_4^{2-}$ in a 0.1040 g sample
   b. 0.228 mol $C_2O_4^{2-}$ and 20.1 g $C_2O_4^{2-}$ in a 100 g sample

4. The $KMnO_4$ solution is such a deep purple that the bottom of the meniscus cannot be seen; therefore the top of the meniscus is read and recorded instead.

5. The reaction of MnO_4^- with $C_2O_4^{2-}$ is slow; the application of some heat ($\approx 60°C$) (1) increases the reaction rate and (2) removes the $CO_2(g)$ as it is formed. In addition the reaction is catalyzed by Mn^{2+}, a reduction product of MnO_4^-; as Mn^{2+} is not initially present in the reaction system the reaction rate is initially slow.

**6. $0.0211 \text{ L} \times \dfrac{0.010 \text{ mol MnO}_4^-}{\text{L}} \times \dfrac{5 \text{ mol C}_2\text{O}_4^{2-}}{2 \text{ mol MnO}_4^-} = 5.28 \times 10^{-4} \text{ mol C}_2\text{O}_4^{2-}$

$0.0147 \text{ g Fe}^{3+} \times \dfrac{\text{mol Fe}^{3+}}{55.85 \text{ g Fe}^{3+}} = 2.63 \times 10^{-4} \text{ mol Fe}^{3+}$

$\dfrac{5.28 \times 10^{-4} \text{ mol C}_2\text{O}_4^{2-}}{2.63 \times 10^{-4} \text{ mol Fe}^{3+}} = 2 \text{ mol C}_2\text{O}_4^{2-} \text{ per 1 mol Fe}^{3+}$

Exclusively of water ligands, the complex must be $[\text{Fe}(\text{C}_2\text{O}_4)_2]^-$; for a neutral salt, one mole of K^+ must be present per mole of complex or the empirical formula of the salt, exclusive of water, is $K[\text{Fe}(\text{C}_2\text{O}_4)_2]$.

Total mass of ions:
$5.28 \times 10^{-4} \text{ mol C}_2\text{O}_4^{2-} = 0.0464 \text{ g C}_2\text{O}_4^{2-}$
$2.63 \times 10^{-4} \text{ mol Fe}^{3+} = 0.0147 \text{ g Fe}^{3+}$
$2.63 \times 10^{-4} \text{ mol K}^+ = 0.0103 \text{ g K}^+$
Mass of water $= 0.100 \text{ g} - (0.0464 \text{ g} + 0.0147 \text{ g} + 0.0103 \text{ g}) = 0.0286 \text{ g H}_2\text{O}$
$0.0286 \text{ g H}_2\text{O} = 1.59 \times 10^{-3} \text{ mol H}_2\text{O}$
$\dfrac{1.59 \times 10^{-3} \text{ mol H}_2\text{O}}{2.63 \times 10^{-4} \text{ mol Fe}^{3+}} = 6.03 \text{ mol H}_2\text{O per 1 mol Fe}^{3+}$

The values of w, x, y and z according to the data are 1:1:2:6 for an empirical formula of $KFe(\text{C}_2\text{O}_4)_2(\text{H}_2\text{O})_6$

PRELABORATORY ASSIGNMENT, EXPERIMENT 39C

1. $A = \log \dfrac{100}{63} = 0.20$

2. $A = a \cdot b \cdot c; \; A = \log \dfrac{100}{48.7} = 0.312$
   $0.312 = 20.83 \text{ L/mol} \cdot \text{cm} \times 1 \text{ cm} \times \text{molar concentration}$
   molar concentration $= 1.5 \times 10^{-2} \text{ mol/L}$

3. a.

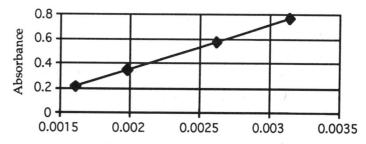

Molar Concentration of Cobalt(II) Ions

   b. $A = \log \dfrac{100}{35.2} = 0.45.$ With $A = 0.45$, then from the graph the molar concentration of cobalt(II) ions in the solution is approximately 2.3×10^{-3} mol/L.

4. 1.22 mL of 0.5 M $CaCl_2$

5. The ascorbic acid reduces all iron(III) ion that is present in the coordination compound to iron(II) ion for the formation of the $[\text{Fe}(\text{bipy})_3]^{2+}$ complex.

LABORATORY QUESTIONS, EXPERIMENT 39A

1. Because iron(II) is oxidized to iron(III), additional oxalate ion is necessary to at least maintain neutrality of the salt; in fact, additional oxalate ion is necessary for the completion of the coordination sphere.

2. Ethanol is less polar than water; the coordination compound in ionic. Being ionic, the coordination compound is less soluble in a less polar solvent and therefore more readily crystallizes.

3. Reducing the volume of the solution increases the concentration of the coordination compound. The intent is that the concentration becomes so high that its maximum solubility is surpassed and the crystals come out of solution.

4. The bidentate oxalate ion binds to the iron(III) ion through its oxygen atoms. In a highly acidic solution, the oxygen atoms become protonated and therefore no longer function as a Lewis base; as a result, the $Fe^{3+}-C_2O_4^{2-}$ Lewis acid-base adduct will not form.

LABORATORY QUESTIONS, EXPERIMENT 39B

1. *Too low.* The moles of oxalate is reported according to the molar concentration of the permanganate solution and its volume. If the molar concentration of the MnO_4^- is reported low, the moles of $C_2O_4^{2-}$ will be reported low.

2. *Too high.* The volume of the MnO_4^- that is delivered from the buret will be recorded as being too high, inferring a greater than actual number of moles of oxalate in the sample.

3. The permanganate ion is an oxidizing agent; the oxalate ion can be oxidized to CO_2, but the Fe^{3+} is already in its highest oxidation state and therefore will not undergo any oxidation.

LABORATORY QUESTIONS, EXPERIMENT 39C

1. Fingerprints on the cuvet reduce the transmission of light through the sample. Higher absorbance data indicates a higher concentration of the *tris*-bipyridyliron(II) ion in the solution and a higher iron(III) ion in the coordination compound.

2. a. *Too high.* A dilution of the solution lowers the molar concentration of the *tris*-bipyridyliron(II) ion and thereby increases the %T of the light.
   b. *Too low.* The dilution also decreases the reported moles of Fe^{3+} in the compound.

3. *Too high.* Some of the Solution 2, which is less concentrated than Solution 3, will dilute the Solution 3. This dilution increases the %T of light for Solution 3.

4. Bipyridyl is a Lewis base. If the pH of the solution is too low, bipyridyl will protonate at the Lewis base sites; as a result the bipyridyl will not form the Lewis acid-base adduct with the iron(II) ion.

5. *Too large.* If the moles of oxalate in a sample is determined to be too high (with the surpassing of the stoichiometric point), and the moles of iron(III) ion is determined accurately, the moles of potassium must be larger for the salt to be neutral. As *w* refers to the moles of potassium, its value will be too large.

LABORATORY QUIZ

1. The absorbance measurement of a 5.0×10^{-5} M $[Fe(bipy)_3]^{2+}$ solution is 0.17 in a 1.5-cm cuvet. What is the molar absorptivity value for $[Fe(bipy)_3]^{2+}$ at this wavelength?
   [Answer: 2.3×10^3 L/mol•cm]

2. What is the purpose of adding ascorbic acid to a sample that contains iron(III) ion for a spectrophotometric analysis of the complex, $[Fe(bipy)_3]^{2+}$?
   [Answer: ascorbic acid reduces Fe(III) to Fe(II)]

3. When measuring the %T of a solution containing $[Fe(bipy)_3]^{2+}$, fingerprints were on the cuvet. How will this affect the reported concentration of iron(II) ion in the sample?
   [Answer: too high]

4. A sample analyzed for oxalate ion required 24.3 mL of 0.0100 M $KMnO_4$ to reach the stoichiometric point in a titration. If 11.31 mg of iron was also know to be in the sample, what is the iron to oxalate mole ratio (in whole numbers) in the sample? [Answer: 1:3]

5. For a 0.100 g sample of $K_wFe_x(C_2O_4)_y(H_2O)_z$, the following data were collected:
   Expt 1: A 23.6-mL volume of 0.010 M $KMnO_4$ was required to titrate the oxalate ion in the sample according to the equation:
   $2 MnO_4^- + 5 H_2C_2O_4 + 6 H^+ \rightarrow 2 Mn^{2+} + 10 CO_2 + 8 H_2O$
   Expt 2: A mass of 11.0 mg of iron(III) was determined to be present.
   What is the formula of the coordination compound?   [Answer: $K_3Fe(C_2O_4)_3(H_2O)_4$

Organic Compounds

INTRODUCTION	The chemical properties of some simple hydrocarbons, alcohols, aldehydes, acids, bases, and esters are studied in this experiment.

The Br_2 and $KMnO_4$ tests are performed on some alkanes, alkenes, alkynes, and aromatic compounds. A simple $K_2Cr_2O_7$ oxidation of primary alcohols (only), a bisulfite test for aldehydes, a litmus test for acids and bases, a $KMnO_4$ test for acids, and the preparation of some "nice-smelling" esters comprise the "chemistry" of the experiment.

WORK ARRANGEMENT	Partners for the tests on the known compounds; individuals for the unknown.

TIME REQUIREMENT	3 hours

LECTURE OUTLINE

1. Follow the Instruction Routine outlined in "To the Laboratory Instructor."

2. Give an overview of functional groups and functional group chemistry as covered in this experiment. The details you present depend upon the extent to which organic chemistry has been covered in lecture.

3. As this is organic chemistry, the use of molecular models in your discussion of functional groups and organic compounds is informative.

4. Discuss the Br_2 and $KMnO_4$ tests.

5. Positive tests are often observed for impure compounds. Check the purity of all test compounds.

6. Advise students of the cautions for the organic chemicals used in the experiment. After study, the organic chemicals should be properly disposed.

7. Students are to design their own Report Sheet for this experiment. Some assistance may be necessary, especially where a structural formula is to be written.

CAUTIONS & DISPOSAL

• Most test compounds are flammable; therefore keep open flames at low heat away from the test compounds and the unknowns.

• Br_2. Bromine can cause severe skin burns

• Oxidizing agents. $KMnO_4$ and $K_2Cr_2O_7$ are strong oxidizing agents. Handle with caution.

• Acids and Bases. Conc H_2SO_4 and glacial CH_3COOH are used as test reagents. Be aware of the techniques used in their handling.

• All test solutions should be discarded in the "Waste Organics" container.

TEACHING HINTS

1. Issue each student an unknown at the beginning of the lab; the test on the unknown can be performed at the same time as that for the known.

2. **Part A.1.** Students should be aware of the potential danger of Br_2; ideally, the tests with the bromine should be done in the hood.

3. **Part B.1.** The $K_2Cr_2O_7$/conc H_2SO_4 combination is strongly oxidizing; handle with caution.

4. **Part D.3.** The oxidation of ethanol to acetic acid is somewhat surprising to students, especially if they are not experienced in organic chemistry.

5. **Part E.** The synthesis of either ester is foolproof; only the odor should be required for a successful synthesis. Some assistance in writing the formula of the ester may be required.

CHEMICALS REQUIRED				
toluene	3 mL	acetaldehyde	1 mL
hexane	0.5 mL	40% $NaHSO_3$	6 mL
1-pentene or 2-pentene	0.5 mL	1 M CH_3COOH	1 mL
1-hexyne	0.5 mL	methylamine	0.5 mL
xylene	0.5 mL	10% $NaHCO_3$	0.5 mL
gasoline	0.5 mL	1% (0.06 M) $KMnO_4$	1 mL
2% Br_2/toluene	3 mL	glacial CH_3COOH	2 mL
acetone	7 mL	conc H_2SO_4	1 mL
1% (0.06 M) $KMnO_4$	2 mL	iso-amyl alcohol	3 mL
0.1 M $K_2Cr_2O_7$	2 mL	salicylic acid	0.5 g
conc H_2SO_4	3 mL	3 M H_2SO_4	0.5 mL
ethanol	10 mL	methanol	0.5 mL

The unknown can be one of the original test samples.

SPECIAL EQUIPMENT			
110°C thermometer	1	ring stand and iron support rings	2
Bunsen burner		cotton plugs	
litmus (red and blue)		"Waste Organics" container	

PRELABORATORY ASSIGNMENT

1. $2 C_8H_{18} + 25 O_2 \rightarrow 16 CO_2 + 18 H_2O$

2. An alcohol can be thought of as a water molecule in which an alkyl group substitutes for an H atom in H–O–H.
CH_3–H (an alkane) CH_3–OH (an alcohol)   H–OH (water)
Alcohols and water are polar and therefore present similar properties.

3. a.   A saturated hydrocarbon is also an alkane, in which all bonds are single bonds; an unsaturated hydrocarbon includes at least one C=C bond or one C≡C bond.
   b.   An alkane has no double bonds; an alkene has at least one C=C bond; an alkyne has at least one C≡C bond.

4. a.   Organic acids can be prepared by the oxidation of an aldehyde.
   b.   An organic acid and an alcohol are necessary for the preparation of an ester.

5. A water molecule becomes an alcohol when an alkyl group is substituted for a hydrogen atom (see answer to Prelaboratory Question 2); an ammonia molecule becomes an amine when an alkyl group is substituted for a hydrogen atom:
H–OH versus CH_3–OH and H–NH_2 versus CH_3–NH_2

6.

a. $H-\overset{\overset{O}{\|}}{C}-OH$ b. $H-\overset{\overset{O}{\|}}{C}-H$ c. $C_3H_7-\overset{\overset{O}{\|}}{C}-O-CH_2CH_3$

LABORATORY QUESTIONS

1. The compound is unsaturated; Br_2 adds across a C=C bond.

2. The compound is unsaturated.

3. a. C_3H_7COOH and C_2H_5OH
   b. CH_3COOH and $C_8H_{17}OH$
   c. $H_2N-C_6H_4COOH$ and CH_3OH

4. a. CH_3–NH_2 is basic to litmus and has an "amine" (fishy) odor. CH_3–OH can be oxidized to formaldehyde, HCHO.

b. CH_3–CHO reacts with $NaHSO_3$, but CH_3OH does not.
c. CH_3COOH is acid to litmus and reacts with $NaHCO_3$ to evolve CO_2
d. CH_3–CH=CH_2 reacts with bromine
e. CH_3–CH=CH_2 reacts with bromine

5. $C_2H_5OH + H_2O \rightarrow CH_3COOH + 4\,H^+ + 4\,e^-$

LABORATORY QUIZ 1. a. Identify a reagent that distinguishes an alkane from an alkene.
b. Identify a reagent that tests for aldehydes.
c. Identify a reagent that tests for organic acids.
d. Identify an oxidizing agent that converts alcohols to organic acids.

2. Write an equation that represents the reaction of Br_2 with propene, $CH_3CH=CH_2$. What evidence indicates that a reaction occurs?

3. What chemical test distinguishes
a. CH_3COOH from $C_2H_5NH_2$?
b. C_2H_6 from C_2H_5OH?
c. C_2H_5OH from $C_2H_5NH_2$?
d. C_2H_6 from $C_2H_5NH_2$?

Reagent Preparations

A large number of different solutions are prepared and stored in the stockroom. Some chemicals and solutions are corrosive, flammable, toxic, incompatible, and unstable. If you are uncertain about the safety precautions for a chemical, consult a safety handbook, the Material Safety Data Sheets (MSDS), or the chemical supply catalog. Become familiar with the NFTA standards which describe the four possible hazards of a chemical and give a numerical rating from 0 to 4. See Technique 3 in the Laboratory Manual.

It is advisable that you always prepare and properly label only *one* solution at a time, using *clean* glassware and equipment. All aqueous solutions should be prepared with (ideally, previously boiled) *deionized* water. Always practice good safety habits—use safety glasses and a lab apron (or coat) and use a fume hood whenever appropriate. Wash your hands frequently, especially after each solution preparation and when leaving the stockroom. *ALWAYS* **PLAY IT SAFE AND** *NEVER* **HURRY.**

The quantity of each chemical listed in each experiment is the amount that is necessary *for each student or group of students*. The stockroom personnel should be aware of the fact that a 10–20% waste allowance must be made when preparing large quantities of solutions for the class.

Acetate buffer solution; $CH_3COOH/NaCH_3CO_2$

0.5 *M*/0.5 *M*   Dissolve 68.04 g $NaCH_3CO_2 \cdot 3H_2O$ in 500 mL deionized water. Add 28.74 mL of glacial acetic acid and dilute to 1 L or mix equal volumes of 1 *M* CH_3COOH and 1 *M* $NaCH_3CO_2$. (Expt 23)

Acetic acid; CH_3COOH, "conc" or glacial is 17.4 *M* (Appendix F, laboratory manual)
Caution: *Skin and mucous irritant. Use adequate ventilation.*

conc	(Expt 40)
0.01 *M*	Dilute 0.58 mL glacial CH_3COOH to volume with water in a 1-L volumetric flask. (Expt 13)
0.05 *M*	Dilute 2.87 mL glacial CH_3COOH to volume with water in a 1-L volumetric flask. (Expt 28)
0.1 *M*	Dilute 5.75 mL glacial CH_3COOH to volume with water in a 1-L volumetric flask. (Expts 13, 24)
0.15 *M*	Dilute 6.82 mL glacial CH_3COOH to volume with water in a 1-L volumetric flask. (Expt 28)
1 *M*	Dilute 57.5 mL glacial CH_3COOH to volume with water in a 1-L volumetric flask. (Expt 40)
3 *M*	Dilute 17.2 mL glacial CH_3COOH to volume with water in a 100-mL volumetric flask. (Expt 13)
6 *M*	Dilute 34.5 mL glacial CH_3COOH to volume with water in a 100-mL volumetric flask. (Expts 13, 22, 34)

Aluminon; molar mass = 473.4 g/mol

0.1%   Dissolve 0.1 g ammonium aurintricarboxylate in 100 mL solution (Expt 35)

Aluminum nitrate; $Al(NO_3)_3 \cdot 9H_2O$, molar mass = 375.14 g/mol

0.1 *M*   Dissolve 37.5 g of $Al(NO_3)_3 \cdot 9H_2O$ with water and dilute to volume with water in a 1-L volumetric flask. (Expt 35)

Ammonia(aq); NH_3, "conc" is 14.8 *M* (Appendix F, laboratory manual)
Caution: *Severe skin and mucous irritant. Use adequate ventilation.*

conc	(Expts 4, 24, 34, 35, 38)
0.10 *M*	Dilute 6.76 mL conc NH_3 to volume with water in a 1-L volumetric flask. (Expt 3)
3 *M*	Dilute 20.2 mL conc NH_3 to volume with water in a 100-mL volumetric flask. (Expt 10)
6 *M*	Dilute 40.5 mL conc NH_3 to volume with water in a 100-mL volumetric flask. (Expts 31, 33, 34, 35, 36)

Ammonium chloride; NH$_4$Cl, molar mass = 53.49 g/mol

0.1 M Dissolve 5.35 g of NH$_4$Cl with water and dilute to volume with water in a 1-L volumetric flask. (Expts 3, 13)

2 M Dissolve 107 g of NH$_4$Cl with water and dilute to volume with water in a 1-L volumetric flask. (Expt 35)

Ammonium molybdate; (NH$_4$)$_6$Mo$_7$O$_{24}$•4H$_2$O, molar mass = 1235.86 g/mol

Caution: *Conc NH$_3$ and conc HNO$_3$ are severe skin and mucous irritants. Prepare Solutions 1 and 2 in the fume hood.*

0.5 M Solution 1: Dissolve 100 g of (NH$_4$)$_6$Mo$_7$O$_{24}$•4H$_2$O in 400 mL of water and 80 mL of conc NH$_3$.

            Solution 2: Mix 400 mL of conc HNO$_3$ with 600 mL of water. Mix, with vigorous stirring, one volume of Solution 1 with 2 volumes of Solution 2 as needed for a laboratory period. *Make up only small amounts–the reagent does not keep.* (Expt 33)

Ammonium nitrate; NH$_4$NO$_3$, molar mass = 80.04 g/mol

Oxidizing salt. Use clean glassware.

0.5 M Dissolve 40.0 g of NH$_4$NO$_3$ with water and dilute to volume with water in a 1-L volumetric flask. (Expt 34)

Ammonium thiocyanate; NH$_4$SCN, molar mass = 76.13 g/mol

0.1 M Dissolve 7.61g of NH$_4$SCN with water and dilute to volume with water in a 1-L volumetric flask. (Expt 35)

Barium chloride; BaCl$_2$•2H$_2$O, molar mass = 244.28 g/mol

0.5 M Dissolve 12.2 g of BaCl$_2$•2H$_2$O with water and dilute to volume with water in a 100-mL volumetric flask. (Expt 8)

Barium nitrate; Ba(NO$_3$)$_2$, molar mass = 261.35 g/mol

0.1 M Dissolve 26.13 g of Ba(NO$_3$)$_2$ with water and dilute to volume with water in a 1-L volumetric flask. (Expts 3, 36)

1 M Dissolve 26.13 g of Ba(NO$_3$)$_2$ with water and dilute to volume with water in a 100-mL volumetric flask. (Expt 33)

Bipyridyl (also, α,α′-dipyridyl or 2,2′-bipyridine); C$_{10}$H$_8$N$_2$, molar mass = 156.19 g/mol

0.02 M Dissolve 0.31 g of C$_{10}$H$_8$N$_2$ with water and dilute to volume with water in a 100-mL volumetric flask. (Expt 39C)

Bismuth nitrate; Bi(NO$_3$)$_3$•5H$_2$O, molar mass = 485.07 g/mol

0.1 M Dissolve 48.5 g of Bi(NO$_3$)$_3$•5H$_2$O with 3 M HNO$_3$ and dilute to volume with 3 M HNO$_3$ in a 1-L volumetric flask. (Expt 34)

Bromine/toluene; Br$_2$/toluene

Caution: *Avoid inhalation or skin contact with Br$_2$ or toluene.*

2%   Add 2 mL of Br$_2$ to 98 mL toluene. Label the solution with appropriate cautions. (Expt 40)

Buffer solution

     Obtain from supplier (Expts 28, 39C)

Calcium chloride; $CaCl_2 \cdot 2H_2O$, molar mass = 147.02 g/mol

0.1 M   Dissolve 14.7 g of $CaCl_2 \cdot 2H_2O$ with water and dilute to volume with water in a 1-L volumetric flask. (Expt 10)

0.5 M   Dissolve 7.35 g of $CaCl_2 \cdot 2H_2O$ with water and dilute to volume with water in a 100-mL volumetric flask. (Expt 39C)

Calcium hydroxide; $Ca(OH)_2$, molar mass = 74.1 g/mol

satd   Mix 2-3 g of $Ca(OH)_2$ with 100 mL of water and let stand for several days. (Expts 3, 30, 33)

satd w/$CaCl_2$ Mix 2-3 g of $Ca(OH)_2$ and 1 g $CaCl_2 \cdot 2H_2O$ with 100 mL of water and let stand for several days. (Expt 30)

Calcium nitrate; $Ca(NO_3)_2 \cdot 4H_2O$, molar mass = 236.16 g/mol

0.3 M   Dissolve 70.8 g of $Ca(NO_3)_2 \cdot 4H_2O$ with water and dilute to volume with water in a 1-L volumetric flask. (Expt 36)

2 M   Dissolve 47.2 g of $Ca(NO_3)_2 \cdot 4H_2O$ with water and dilute to volume with water in a 100-mL volumetric flask. (Expt 10)

Calcium oxide (quicklime); CaO

Heat $Ca(OH)_2$ to remove water and cool in a desiccator. (Expt 13)

Cobalt(II) chloride; $CoCl_2 \cdot 6H_2O$, molar mass = 237.93 g/mol
Avoid skin contact.

0.1 M   Dissolve 23.8 g of $CoCl_2 \cdot 6H_2O$ with water and dilute to volume with water in a 1-L volumetric flask. (Expt 38)

1 M   Dissolve 23.8 g of $CoCl_2 \cdot 6H_2O$ with water and dilute to volume with water in a 100-mL volumetric flask. (Expt 24)

Cobalt(II) nitrate; $Co(NO_3)_2 \cdot 6H_2O$, molar mass = 291.04 g/mol

0.1 M   Dissolve 29.1 g of $Co(NO_3)_2 \cdot 6H_2O$ with water and dilute to volume with water in a 1-L volumetric flask. (Expt 38)

0.1 M   Dissolve 2.91 g of $Co(NO_3)_2 \cdot 6H_2O$ with 0.1 M HNO_3 and dilute to volume with 0.1 M HNO_3 in a 100-mL volumetric flask. (Expt 4)

Copper(II) nitrate; $Cu(NO_3)_2 \cdot 3H_2O$, molar mass = 241.60 g/mol

0.1 M   Dissolve 24.2 g of $Cu(NO_3)_2 \cdot 3H_2O$ with water and dilute to volume with water in a 1-L volumetric flask. (Expts 14, 31, 34, 38)

0.1 M   Dissolve 2.42 g of $Cu(NO_3)_2 \cdot 3H_2O$ with 0.1 M HNO_3 and dilute to volume with 0.1 M HNO_3 in a 100-mL volumetric flask. (Expt 4)

1 M   Dissolve 24.2 g of $Cu(NO_3)_2 \cdot 3H_2O$ with water and dilute to volume with water in a 100-mL volumetric flask. (Expt 33)

Copper(II) sulfate; $CuSO_4 \cdot 5H_2O$, molar mass = 249.68 g/mol
Caution: *Slowly add the conc H_2SO_4 to a large quantity of water.*

0.001 M   Dissolve 0.250 g of $CuSO_4 \cdot 5H_2O$ with water, add 5 mL of conc H_2SO_4, and dilute to volume with water in a 1-L volumetric flask. (Expt 31)

0.1 M   Dissolve 25.0 g of $CuSO_4 \cdot 5H_2O$ with water, add 5 mL of conc H_2SO_4, and dilute to volume with water in a 1-L volumetric flask. (Expts 3, 24, 32, 38)

1 M   Dissolve 25.0 g of $CuSO_4 \cdot 5H_2O$ with water, add 0.5 mL of conc H_2SO_4, and dilute to volume with water in a 100-mL volumetric flask. (Expts 31, 32)

Dimethylglyoxime; HON=C(CH₃)C(CH₃)=NOH (H₂DMG), molar mass = 116.12 g/mol

0.1 *M* (or 1%)   Dissolve 1.2 g of $HON=C(CH_3)C(CH_3)=NOH$ (H_2DMG) in 100 mL of 95% ethanol. (Expts 4, 35)

Ethanol

50%   (Expt 37)
75%   (Expt 39A)

Hydrochloric acid; HCl(aq), "conc" is 12.1 M (Appendix F, laboratory manual)
Caution: *Severe skin and mucous irritant. Use adequate ventilation.*
Caution: *Always add the concentrated acid to water, never vice versa.*

conc   (Expts 11, 13, 24, 36, 38)
0.00010 *M*   Dilute 1.0 mL of 0.1 *M* HCl with water and dilute to volume with water in a 1-L volumetric flask. (Expt 13)
0.010 *M*   Dilute 10 mL of 0.1 *M* HCl with water and dilute to volume with water in a 100-mL volumetric flask. (Expt 13)
0.05 *M*   (Standardize) Prepare a 0.05 *M* HCl solution and standardize against anhydrous Na_2CO_3. Weigh ≈ 0.05-g samples of *dry* Na_2CO_3. Dissolve with water and titrate with the acid, using a methyl purple or bromocresol green indicator. At the first color change, heat the solution to boiling for 2–3 minutes to expel CO_2. Consult an analytical text for a detailed procedure. (Expt 30)
0.10 *M*   Dilute 8.3 mL of conc HCl to volume with water in a 1-L volumetric flask. *Add some water to the volumetric flask first!* (Expts 3, 9, 13, 22, 24, 34)
0.1 *M*   (Standardize) Prepare a 0.1 *M* HCl solution and standardize against anhydrous Na_2CO_3. Weigh 0.1-g to 0.15-g samples of *dry* Na_2CO_3. Dissolve with water and titrate with the acid, using a methyl purple or bromocresol green indicator. At the first color change, heat the solution to boiling for 2 to 3 minutes to expel CO_2. Consult an analytical text for a detailed procedure. (Expt 26)
0.15 *M*   Dilute 12.4 mL of conc HCl to volume with water in a 1-L volumetric flask. *Add some water to the volumetric flask first!* (Expt 9)
0.20 *M*   Dilute 16.5 mL of conc HCl to volume with water in a 1-L volumetric flask. *Add some water to the volumetric flask first!* (Expt 9)
1.0 *M*   Dilute 82.6 mL of conc HCl to volume with water in a 1-L volumetric flask. *Add some water to the volumetric flask first!* (Expts 6, 7, 22, 24)
1.1 *M*   Dilute 90.9 mL of conc HCl to volume with water in a 1-L volumetric flask. *Add some water to the volumetric flask first!* (Expt 21)
3.0 *M*   Dilute 24.8 mL of conc HCl to volume with water in a 100-mL volumetric flask. *Add some water to the volumetric flask first!* (Expts 10, 12, 13, 18, 22)
6.0 *M*   Dilute 49.6 mL of conc HCl to volume with water in a 100-mL volumetric flask. *Add some water to the volumetric flask first!* (Expts 4, 10, 13, 14, 19, 22, 34, 35, 36)

Hydrogen peroxide; H₂O₂
Use clean glassware

3%   Purchase from chemical supplier. (Expt 22)
0.1 *M* Dilute 113 mL of 3% H_2O_2 to volume with water in a 1-L volumetric flask. (Expt 23)
6%   Dilute 20.0 mL of 30% H_2O_2 to volume with water in a 100-mL volumetric flask. (Expt 39A)

Hydroxylammonium chloride; NH₃OH⁺Cl⁻ (also, hydroxylamine hydrochloride; NH₂OH•HCl), molar mass = 69.49 g/mol

0.05 *M*   Dissolve 0.347 g of $NH_3OH^+Cl^-$ with water and dilute to volume in a 100-mL volumetric flask. (Expt 16)

Iodic acid; HIO_3, molar mass = 175.91 g/mol
Avoid skin contact.

0.01 M   Dissolve 1.76 g of HIO_3 with water and dilute to volume with water in a 1-L volumetric flask *or* mix equal volumes of 0.02 M KIO_3 and 0.02 M H_2SO_4 prior to its use. The solution is stable for about 48 hrs. (Expt 22)

Iron(II) or ferrous ammonium sulfate; $Fe(NH_4)_2(SO_4)_2 \cdot 6H_2O$, molar mass = 392.15 g/mol
Prepare fresh solutions for best results. Fe^{2+} air oxidizes to Fe^{3+}.
Caution: *Be careful in handling the conc H_2SO_4.*

0.1 M   Dissolve 39.2 g of $Fe(NH_4)_2(SO_4)_2 \cdot 6H_2O$ with water and dilute to volume with water containing 2 mL of conc H_2SO_4 in a 1 L volumetric flask. (Expts 14, 31)

Iron(III) or ferric ammonium sulfate; $FeNH_4(SO_4)_2 \cdot 12H_2O$, molar mass = 482.19 g/mol

0.1 M   Dissolve 48.2 g of $FeNH_4(SO_4)_2 \cdot 12H_2O$ with water and dilute to volume with water containing 1 mL of conc H_2SO_4 in a 100-mL volumetric flask. (Expt 4)

0.2 M   Dissolve 96.4 g of $FeNH_4(SO_4)_2 \cdot 12H_2O$ with water and dilute to volume with water containing 1 mL of conc H_2SO_4 in a 100-mL volumetric flask. (Expt 16)

Iron(III) or ferric nitrate; $Fe(NO_3)_3 \cdot 9H_2O$, molar mass = 404.00 g/mol

0.002 M   Dissolve 0.808 g of $Fe(NO_3)_3 \cdot 9H_2O$ with 0.25 M HNO_3 and dilute to volume with 0.25 M HNO_3 in a 1-L volumetric flask. (Expt 25)

0.1 M   Dissolve 40.4 g of $Fe(NO_3)_3 \cdot 9H_2O$ with 0.25 M HNO_3 and dilute to volume with 0.25 M HNO_3 in a 1-L volumetric flask. (Expts 10, 35)

0.2 M   Dissolve 80.8 g of $Fe(NO_3)_3 \cdot 9H_2O$ with 0.25 M HNO_3 and dilute to volume with 0.25 M HNO_3 in a 1-L volumetric flask. (Expts 25, 33)

Iron(II) or ferrous sulfate, $FeSO_4 \cdot 7H_2O$, molar mass = 278.03 g/mol
Prepare fresh solutions for best results. Fe^{2+} air oxidizes to Fe^{3+}.

0.1 M   Dissolve 27.8 g of $FeSO_4 \cdot 7H_2O$ with water and dilute to volume with water containing 10 mL of conc H_2SO_4 in a 1-L volumetric flask. Add a small piece of clean iron wire to inhibit Fe^{2+} oxidation. (Expt 31)

satd   Add about 20 g of $FeSO_4 \cdot 7H_2O$ to 100 mL of water. (Expt 33)

Lead(II) nitrate; $Pb(NO_3)_2$, molar mass = 331.23 g/mol
Avoid skin contact. Dispose of the excess solution in a "Waste Metal Salts" container.

0.1 M   Dissolve 3.31 g of $Pb(NO_3)_2$ with water and dilute to volume with water in a 100 mL volumetric flask. (Expt 31)

Limewater, saturated $Ca(OH)_2$ solution

See calcium hydroxide (Expt 33)

Magnesium chloride; $MgCl_2 \cdot 6H_2O$, molar mass = 203.30 g/mol

0.10 M   Dissolve 20.3 g of $MgCl_2 \cdot 6H_2O$ with water and dilute to volume with water in a 1-L volumetric flask. (Expt 10)

Magnesium nitrate; $Mg(NO_3)_2 \cdot 6H_2O$, molar mass = 256.41 g/mol

0.1 M   Dissolve 25.6 g of $Mg(NO_3)_2 \cdot 6H_2O$ with water and dilute to volume with water in a 1-L volumetric flask. (Expts 31, 36)

Magnesium sulfate; MgSO₄•7H₂O, molar mass = 246.48 g/mol

$0.1\ M$ Dissolve 2.46 g MgSO₄•7H₂O with water and dilute to volume in a 100-mL volumetric flask. (Expt 3)

Manganese(II) chloride MnCl₂•4H₂O, molar mass = 197.91 g/mol

$0.1\ M$ Dissolve 19.8 g of MnCl₂•4H₂O with water and dilute to volume with water in a 1-L volumetric flask. (Expt 35)

Manganese(II) nitrate; 50% Mn(NO₃)₂, d = 1.6 g/mL

$0.1\ M$ Dilute 22.4 mL of 50% Mn(NO₃)₂ with water to volume in a 1-L volumetric flask. (Expt 35)

$1\ M$ Dilute 22.4 mL of 50% Mn(NO₃)₂ with 0.1 M HNO₃ to volume in a 100-mL volumetric flask. (Expt 4)

Methylammonium chloride (or methylamine hydrochloride); CH₃NH₂•HCl, molar mass = 67.52 g/mol

$0.1\ M$ Dissolve 6.75 g of CH₃NH₂•HCl with water and dilute to volume with water in a 1-L volumetric flask. (Expt 3)

Nickel(II) chloride; NiCl₂•6H₂O, molar mass = 237.71 g/mol

$0.1\ M$ Dissolve 23.7 g NiCl₂•6H₂O with water and dilute to volume with water in a 1-L volumetric flask. (Expts 24, 38)

Nickel(II) nitrate; Ni(NO₃)₂•6H₂O, molar mass = 290.81 g/mol

$0.1\ M$ Dissolve 29.1 g of Ni(NO₃)₂•6H₂O with water and dilute to volume with water in a 1-L volumetric flask. (Expts 31, 35, 38)

$1\ M$ Dissolve 29.1 g of Ni(NO₃)₂•6H₂O with 0.1 M HNO₃ and dilute to volume with 0.1 M HNO₃ in a 100-mL volumetric flask. (Expt 4)

Nickel(II) sulfate; NiSO₄•6H₂O, molar mass 262.86 g/mol

$0.1\ M$ Dissolve 26.2 g of NiSO₄•6H₂O with water and dilute to volume with water in a 1-L volumetric flask. (Expt 14)

Nitric acid; HNO₃, "conc" is 15.7 M, (Appendix F, laboratory manual)

Caution: *Handle with care; conc HNO₃ causes severe skin burns.*
Caution: *HNO₃ is an oxidizing acid. Use clean glassware and use adequate ventilation.*
Caution: *Always the concentrated acid to water or to a more dilute solution of nitric acid.*

conc   (Expts 5, 14, 35, 37)

$0.1\ M$ Dilute 6.37 mL of conc HNO₃ to volume with water in a 1-L volumetric flask. *Add some water to the volumetric flask first!* (Expts 3, 14)

$0.25\ M$ Dilute 15.9 mL of conc HNO₃ to volume with water in a 1-L volumetric flask. *Add some water to the volumetric flask first!* (Expt 25)

$1\ M$ Dilute 63.7 mL of conc HNO₃ to volume with water in a 1-L volumetric flask. *Add some water to the volumetric flask first!* (Expts 31, 32)

$1.1\ M$ Dilute 70.1 mL of conc HNO₃ to volume with water in a 1-L volumetric flask. *Add some water to the volumetric flask first!* (Expt 21)

$6\ M$ Dilute 38.2 mL of conc HNO₃ to volume with water in a 100-mL volumetric flask. *Add some water to the volumetric flask first!* (Expts 6, 7, 10, 13, 14, 24, 33, 34, 35)

Oxalic acid; $H_2C_2O_4 \cdot 2H_2O$, molar mass = 126.1 g/mol
Avoid skin contact.

0.33 *M*   Dissolve 41.6 g $H_2C_2O_4 \cdot 2H_2O$ with water and dilute to volume with water in a 1-L volumetric flask. (Expt 22)

1 *M*   Dissolve 12.6 g $H_2C_2O_4 \cdot 2H_2O$ with water and dilute to volume with water in a 100-mL volumetric flask. (Expt 39A)

Phosphoric acid; H_3PO_4, "conc" is 85% H_3PO_4, d = 1.685 g/mL (Appendix F, laboratory manual)
Avoid skin contact. Always add the conc acid to the water in the dilution of an acid.

conc   (Expt 13)

0.1 *M*   Dilute 6.80 mL of conc H_3PO_4 to volume with water in a 1-L volumetric flask. *Add some water to the volumetric flask first!* (Expt 3)

6 *M*   Dilute 40.8 mL of conc H_3PO_4 to volume with water in a 100-mL volumetric flask. *Add some water to the volumetric flask first!* (Expts 16, 22)

Potassium aluminum sulfate; $KAl(SO_4)_2 \cdot 12H_2O$, molar mass = 474.39 g/mol

0.1 *M*   Dissolve 47.4 g with water and dilute to volume with water in a 100-mL volumetric flask. (Expt 13)

Potassium bromide, KBr, molar mass = 119.00 g/mol

6 *M*   Dissolve 71.4 g of KBr with water and dilute to volume with water in a 100-mL volumetric flask. (Expt 10)

Potassium chromate; K_2CrO_4, molar mass = 194.20 g/mol
Caution: A strong oxidant. Avoid skin contact. Dispose of the excess solution in a "Waste Metal Salts" container.

1 *M*   Dissolve 19.4 g of K_2CrO_4 with water and dilute to volume with water in a 100-mL volumetric flask. (Expt 36)

Potassium dichromate; $K_2Cr_2O_7$, molar mass = 294.19 g/mol
Caution: A strong oxidant. Avoid skin contact. Dispose of the excess solution in a "Waste Metal Salts" container.

0.01 *M*   Dissolve 2.94 g of $K_2Cr_2O_7$ with water and dilute to volume with water in a 1-L volumetric flask. (Expt 14)

0.1 *M*   Dissolve 29.4 g of $K_2Cr_2O_7$ with water and dilute to volume with water in a 1-L volumetric flask. (Expt 40)

Potassium hexacyanoferrate(II); $K_4[Fe(CN)_6] \cdot 3H_2O$, molar mass = 422.41 g/mol

0.2 *M*   Dissolve 84.5 g of $K_4[Fe(CN)_6] \cdot 3H_2O$ with water and dilute to volume with water in a 1-L volumetric flask. (Expts 4, 34, 35)

Potassium hydroxide; KOH, molar mass = 56.11 g/mol
Caution: The dissolution of KOH in water is very exothermic.

4 *M*   Dissolve 22.4 g of KOH with water and dilute to volume with water in a 100-mL volumetric flask. (Expt 37)

Potassium iodide; KI, molar mass = 166.01 g/mol

0.1 *M*   Dissolve 16.6 g of KI with water and dilute to volume with water in a 1-L volumetric flask. (Expts 14, 24)

0.3 *M*   Dissolve 4.98 g of KI with water and dilute to volume with water in a 100-mL volumetric flask. (Expt 23)

| 6 *M* | Dissolve 99.6 g of KI with water and dilute to volume with water in a 100-mL volumetric flask. (Expt 10) |

Potassium nitrate, KNO₃, molar mass = 101.11 g/mol
An oxidizing salt. Use clean glassware.

| 0.1 *M* | Dissolve 10.1 g of KNO_3 with water and dilute to volume with water in a 1-L volumetric flask. (Expts 3, 31) |
| 0.25 *M* | Dissolve 2.53 g of KNO_3 with water and dilute to volume with water in a 100-mL volumetric flask. (Expt 34) |

Potassium oxalate; K₂C₂O₄•H₂O, molar mass = 184.24 g/mol

| 1 *M* | Dissolve 18.4 g of $K_2C_2O_4 \cdot H_2O$ with water and dilute to volume with water in a 100-mL volumetric flask. (Expts 14, 36) |

Potassium permanganate, KMnO₄, molar mass = 158.04 g/mol
Caution: *A strong oxidant. Store in a clean, amber bottle (KMnO₄ decomposes in light.)*

0.01 *M*	Dissolve 1.58 g of $KMnO_4$ with water and dilute to volume with water in a 1-L volumetric flask. (Expt 14)
0.01 *M*	Dissolve 1.58 g of $KMnO_4$ with 3 *M* H_2SO_4 and dilute to volume with 3 *M* H_2SO_4 in a 1-L volumetric flask. (Expt 22)
0.01 *M*	(Standardized) Dissolve 1.58 g of $KMnO_4$ with previously boiled, deionized water and dilute to volume with the same water in a 1-L volumetric flask. Let stand one day. Standardize against dried 0.05-g samples of $Na_2C_2O_4$. Measure 0.05-g samples of *dry* $Na_2C_2O_4$, dissolve in 100 mL of 1 *M* H_2SO_4, and heat nearly to boiling. Titrate at temperatures ≥70°C, with the $KMnO_4$ solution until the pink coloration persists. Refer to an analytical text for the experimental procedure and techniques. (Expt 39B)
1% (0.06 *M*)	Dissolve 1 g of $KMnO_4$ with 99g of water. (Expt 40)
0.1 *M*	Dissolve 1.58 g of $KMnO_4$ with water and dilute to volume with water in a 100-mL volumetric flask. (Expt 33)

Potassium thiocyanate; KSCN, molar mass = 97.18 g/mol

0.1 *M*	Dissolve 0.972 g KSCN with water and dilute to volume with water in a 100-mL volumetric flask. (Expt 38)
0.2 *M*	Dissolve 1.94 g KSCN with water and dilute to volume with water in a 100-mL volumetric flask. (Expt 4)
satd in acetone	Place 2–3 g of KSCN in 50 mL of acetone and store in a dropper bottle. (Expt 4)

Silver nitrate; AgNO₃, molar mass = 169.87 g/mol
Avoid skin contact. Store in an amber bottle.
Avoid skin contact. Dispose of the excess solution in a "Waste Metal Salts" container.

| 0.01 *M* | Dissolve 0.170 g of $AgNO_3$ with water and dilute to volume with water in a 100-mL volumetric flask. (Expts 24, 33) |
| 0.1 *M* | Dissolve 1.70 g of $AgNO_3$ with water and dilute to volume with water in a 100-mL volumetric flask. (Expts 3, 10, 31, 34) |

Silver sulfate; Ag₂SO₄, molar mass = 311.80 g/mol
Avoid skin contact. Store in an amber bottle.
Avoid skin contact. Dispose of the excess solution in a "Waste Metal Salts" container.

| 0.04 *M* | (Saturated) Dissolve 1.25 g of Ag_2SO_4 with water and dilute to volume with water in a 100-mL volumetric flask. (Expt 33) |

Sodium acetate; NaCH₃CO₂•3H₂O, molar mass = 136.08 g/mol or NaCH₃CO₂, molar mass = 82.03 g/mol

0.1 M Dissolve 13.6 g of NaCH₃CO₂•3H₂O with water and dilute to volume with water in a 1-L volumetric flask *or* dissolve 8.20 g of NaCH₃CO₂ in water and dilute to 1-L in a volumetric flask. (Expt 24)

Sodium bismuthate; NaBiO₃, molar mass = 279.97 g/mol
Caution: *strong oxidizing agent; moisture sensitive.*

0.1 M Dissolve 2.8 g of NaBiO₃ with water and dilute to volume with water in a 100-mL volumetric flask. (Expt 4)

Sodium bromide; NaBr, molar mass = 102.89 g/mol

0.2 M Dissolve 20.6 g of NaBr with water and dilute to volume with water in a 1-L volumetric flask. (Expt 33)

Sodium carbonate; Na₂CO₃, molar mass = 105.99 g/mol

0.1 M Dissolve 10.6 g of Na₂CO₃ with water and dilute to volume with water in a 1-L volumetric flask. (Expts 3, 13, 24)
0.2 M Dissolve 21.2 g of Na₂CO₃ with water and dilute to volume with water in a 1-L volumetric flask. (Expt 33)

Sodium chloride; NaCl, molar mass = 58.44 g/mol

0.1 M Dissolve 5.84 g of NaCl with water and dilute to volume with water in a 1-L volumetric flask. (Expts 3, 13)
0.2 M Dissolve 11.7 g of NaCl with water and dilute to volume with water in a 1-L volumetric flask. (Expt 33)

Sodium dihydrogen phosphate (also, sodium phosphate, monobasic); NaH₂PO₄, molar mass = 120.05 g/mol or NaH₂PO₄•H₂O, molar mass = 137.99 g/mol

0.1 M Dissolve 12.0 g of NaH₂PO₄ in water *or* 13.8 g of NaH₂PO₄•H₂O with water and dilute to volume with water in a 1-L volumetric flask. (Expt 13)

Sodium hydrogen carbonate (sodium bicarbonate); NaHCO₃, molar mass = 84.01g/mol

10%   Dissolve 10 g of NaHCO₃ with 90 g of water. (Expt 40)

Sodium (mono)hydrogen phosphate (also, sodium phosphate, dibasic); Na₂HPO₄•12H₂O, molar mass = 358.14 g/mol or Na₂HPO₄•7H₂O, molar mass = 268.07 g/mol

1 M Dissolve 35.8 g of Na₂HPO₄•12H₂O *or* 26.8g Na₂HPO₄•7H₂O with water and dilute to volume with water in a 100-mL volumetric flask. (Expt 36)

Sodium hydrogen sulfite (sodium bisulfite); NaHSO₃•H₂O, molar mass = 104.06 g/mol

0.1 M Dissolve 10.4 g of NaHSO₃•H₂O with water and dilute to volume with water in a 1-L volumetric flask. (Expt 14)
40%   Dissolve 48.4 g NaHSO₃•H₂O in 51.6 g of water. (Expt 40)

Sodium hydroxide; NaOH, molar mass = 40.00 g/mol

Caution: *Add solid NaOH slowly with stirring–the dissolution is exothermic.
Store all NaOH solutions in a polyethylene container.*

conc	(Expt 9)
0.00010 M	Dilute 1 mL of 0.1 M NaOH to volume with (boiled, deionized) water in a 1-L volumetric flask. (Expt 13)
0.010 M	Dissolve 0.40 g of NaOH with (boiled, deionized) water and dilute to volume with the same water in a 1-L volumetric flask. (Expt 10)
0.10 M	Dissolve 4.0 g of NaOH with (boiled, deionized) water and dilute to volume with the same water in a 1-L volumetric flask. (Expts 3, 13, 24).
0.10 M	(Standardize) Prepare a 0.1 M NaOH solution. Cap tightly. Standardize against dried, 2-g samples of potassium hydrogen phthalate, $KHC_8H_4O_4$, molar mass = 204.2 g/mol. Refer to Experiment 9 ($\approx$0.15 M NaOH) in the laboratory manual or in an analytical text for the experimental procedure and techniques. (Expts 26, 27, 28, 29)
0.2 M	Dissolve 8.0 g of NaOH with (boiled, deionized) water and dilute to volume with the same water in a 1-L volumetric flask. (Expt 13)
1.0 M	Dissolve 40.0 g of NaOH with (boiled, deionized) water and dilute to volume with the same water in a 1-L volumetric flask. (Expts 13, 38)
1.0 M	(Standardize) Dissolve 40.0 g of NaOH with (boiled, deionized) water and dilute to volume with the same water in a 1-L volumetric flask. Cap tightly. Standardize against dried, 2-g samples of potassium hydrogen phthalate, $KHC_8H_4O_4$, molar mass = 204.2 g/mol. Refer to Experiment 9 in the laboratory manual or an analytical text for the experimental procedure and techniques. (Expt 21)
3.0 M	Dissolve 12.0 g of NaOH with (boiled, deionized) water and dilute to volume with the same water in a 100-mL volumetric flask. (Expts 12, 33)
6.0 M	Dissolve 24.0 g of NaOH with (boiled, deionized) water and dilute to volume with the same water in a 100-mL volumetric flask. (Expts 5, 10, 34, 35)

Sodium hypochlorite; NaOCl, molar mass = 74.44 g/mol

5%	Purchase as 5% available chlorine solution. Also, a bleach solution purchased from the supermarket is sufficient for these experiments. (Expts 14, 15)

Sodium iodide; NaI, molar mass = 149.89 g/mol

0.2 M	Dissolve 30.0 g of NaI with water and dilute to volume with water in a 1-L volumetric flask. (Expt 33)

Sodium nitrate; NaNO3, molar mass = 84.99 g/mol

An oxidizing salt. Use clean glassware.

0.20 M	Dissolve 17.0 g of $NaNO_3$ with water and dilute to volume with water in a 1-L volumetric flask. (Expt 33)
0.25 M	Dissolve 21.2 g of $NaNO_3$ with water and dilute to volume with water in a 1-L volumetric flask. (Expt 34)

Sodium phosphate (also, sodium phosphate, tribasic); Na3PO4, molar mass = 163.94 g/mol or Na3PO4•12H2O, molar mass = 380.12 g/mol

0.10 M	Dissolve 16.4 g of Na_3PO_4 in water or dissolve 38.0 g of $Na_3PO_4 \cdot 12H_2O$ with water and dilute to volume with water in a 1-L volumetric flask. (Expts 3, 13)
0.20 M	Dissolve 32.8 g of Na_3PO_4 in water or dissolve 76.0 g of $Na_3PO_4 \cdot 12H_2O$ with water and dilute to volume with water in a 1-L volumetric flask. (Expt 33)
0.50 M	Dissolve 8.19 g of Na_3PO_4 in water or dissolve 19.0 g of $Na_3PO_4 \cdot 12H_2O$ with water and dilute to volume with water in a 100-mL volumetric flask. (Expt 8)

Sodium sulfate; Na_2SO_4, molar mass = 142.04 g/mol

0.10 M	Dissolve 14.2 g of Na_2SO_4 with water and dilute to volume with water in a 1-L volumetric flask. (Expt 3)
0.20 M	Dissolve 28.4 g of Na_2SO_4 with water and dilute to volume with water in a 1-L volumetric flask. (Expt 33)

Sodium sulfide; $Na_2S \cdot 9H_2O$, molar mass = 240.18 g/mol or Na_2S, molar mass = 78.04 g/mol

0.10 M	Dissolve 7.80 g of Na_2S in water *or* dissolve 24.0 g of $Na_2S \cdot 9H_2O$ with water and dilute to volume with water in a 1-L volumetric flask. (Expts 3, 24)
0.2 M	Dissolve 15.6 g of Na_2S in water *or* dissolve 48.0 g of $Na_2S \cdot 9H_2O$ with water and dilute to volume with water in a 1-L volumetric flask. (Expt 33)

Sodium thiocyanate; NaSCN, molar mass = 81.08 g/mol

0.002 M	Dissolve 0.162 (±0.001 g) of NaSCN with 0.25 M HNO_3 and dilute to volume with 0.25 M HNO_3 in a 1-L volumetric flask. (Expt 25)

Sodium thiosulfate; $Na_2S_2O_3 \cdot 5H_2O$, molar mass = 248.18 g/mol

0.02 M	Dissolve 4.96 g of $Na_2S_2O_3 \cdot 5H_2O$ with (boiled, distilled) water and dilute to volume with the same water in a 1-L volumetric flask. (Expt 23)
0.10 M	Dissolve 24.8 g of $Na_2S_2O_3 \cdot 5H_2O$ with (boiled, distilled) water and dilute to volume with the same water in a 1-L volumetric flask. (Expts 15, 22)
0.10 M	(Standardized) Dissolve 24.8 g of $Na_2S_2O_3 \cdot 5H_2O$ with (boiled, distilled) water and dilute to volume with the same water in a 1-L volumetric flask. Standardize this solution against 0.2-g samples of *dry* $K_2Cr_2O_7$ dissolved in 100 mL of water and 4 mL of conc H_2SO_4 (**Caution!**). Add 2 g Na_2CO_3, 5 g KI/5 mL water, swirl, cover, and let stand for 3 min. Dilute to about 200 mL and titrate with the thiosulfate solution until the yellow I_2 nearly disappears. Add 5 mL of starch solution and continue titrating until the blue color disappears. See a standard analytical laboratory text for a more complete procedure. (Expt 15)

Starch

1%	Prepare a *paste* with 1 g soluble potato starch and 10 to 20 mL of water. Add the paste to 100 mL of boiling water and boil for 3 minutes. Cool and store in a clean, glass-stoppered bottle. The shelf life is approximately one week. (Expts 14, 15, 22, 23)

Sulfuric acid; H_2SO_4,"conc" is 17.8 M (Appendix F, laboratory manual)

Caution: *Conc H_2SO_4 is an oxidizing acid and a severe skin and mucous irritant. Use clean glassware and adequate ventilation.*
Caution: *Add the conc H_2SO_4 slowly, with stirring to water–the mixing and dissolution process is very exothermic. Cool as necessary. Avoid skin contact.*

conc	(Expts 10, 13, 29, 33, 40)
0.10 M	Dilute 5.62 mL of conc H_2SO_4 to volume with water in a 1-L volumetric flask. *Add some water to the volumetric flask first!* (Expts 10, 13)
0.50 M	Dilute 28.1 mL of conc H_2SO_4 to volume with water in a 1-L volumetric flask. *Add some water to the volumetric flask first!* (Expt 15)
0.9 M	Dilute 50.6 mL of conc H_2SO_4 to volume with water in a 1-L volumetric flask. *Add some water to the volumetric flask first!* (Expt 16)
2.0 M	Dilute 112 mL of conc H_2SO_4 to volume with water in a 1-L volumetric flask. *Add some water to the volumetric flask first!* (Expt 37)
3.0 M	Dilute 16.9 mL of conc H_2SO_4 to volume with water in a 100-mL volumetric flask. *Add some water to the volumetric flask first!* (Expts 22, 40)
6.0 M	Dilute 33.7 mL of conc H_2SO_4 (*slowly*) to volume with water in a 100-mL volumetric flask. *Add some water to the volumetric flask first!* (Expts 5, 14, 36, 37, 39A, 39B, 39C)

Sulfurous acid; H_2SO_3

0.01 M Mix equal volumes of 0.02 M Na_2SO_3 (molar mass = 126 g/mol) and 0.02 M H_2SO_4 just prior to the laboratory period. This mixture should be checked for effectiveness after 10 hours. (Expt 22)

Thioacetamide; CH_3CSNH_2, molar mass = 75.13 g/mol

1.0 M Dissolve 7.5 g of CH_3CSNH_2 with water and dilute to volume with water in a 100-mL volumetric flask. Prepare in small quantities as needed. (Expts 34, 35)

Tin(II) chloride; $SnCl_2 \cdot 2H_2O$, molar mass = 225.63 g/mol

Dissolution of the tin(II) chloride may require some heating.

0.10 M Dissolve 22.5 g of $SnCl_2 \cdot 2H_2O$ with 3 M HCl and dilute to volume with 3 M HCl in a 1-L volumetric flask. Add a few pieces of tin metal. (Expt 31)

1.0 M Dissolve 22.5 g of $SnCl_2 \cdot 2H_2O$ with 3 M HCl and dilute to volume with 3 M HCl in a 100-mL volumetric flask. Add a few pieces of tin metal. (Expt 34)

Zinc nitrate; $Zn(NO_3)_2 \cdot 6H_2O$, molar mass = 297.47 g/mol

0.10 M Dissolve 29.7 g of $Zn(NO_3)_2 \cdot 6H_2O$ with water and dilute to volume with water in a 1 L volumetric flask. (Expts 14, 31, 35)

NOTES FOR REAGENT PREPARATIONS

Preparation of Indicators

	pH Range	Color Change

Alizarin yellow R

Dissolve 0.01 g alizarin yellow R (5-(*p*-nitrophenylazo)salicylic acid, Na-salt) in 100 mL H_2O.　10.1–12.0　yellow to red

Bromocresol purple

Dissolve 0.1 g bromocresol purple (5',5"-dibromo-*o*-cresolsulfone-phthalein) in 18.5 mL of 0.01 *M* NaOH and dilute to 250 mL with H_2O.　5.2–6.8　yellow to purple

Bromophenol blue

Mix 0.1 g bromophenol blue (3', 3", 5', 5"-tetrabromophenol-sulfonephthalein) and 14.9 mL of 0.01 *M* NaOH. Dilute to 250 mL with H_2O. If purchased as the sodium salt, dissolve 0.1 g in 100 mL solution. Prepared indicator can be purchased. (Expt 26)　3.0–4.6　yellow to blue

Bromothymol blue

Mix 0.1 g bromothymol blue (3', 3"-dibromothymolsulfone-phthalein) and 16 mL of 0.01 *M* NaOH. Dilute to 250 mL with H_2O. If purchased as the sodium salt, dissolve 0.1 g in 100 mL solution. Prepared indicator can be purchased.　6.0–7.6　yellow to blue

Methyl orange

Dissolve 0.01 g methyl orange in 100 mL H_2O. Filter if necessary. (Expt 30)　3.2–4.4　red to yellow

Methyl red

Dissolve 0.02 g methyl red in 60.0 mL ethanol and dilute with 40.0 mL H_2O.　4.8–6.0　red to yellow

p-Nitrophenol

Dissolve 0.1 g *p*-nitrophenol in 100 mL H_2O.　5.4–6.6　colorless to yellow

Phenolphthalein

Dissolve 0.05 g phenolphthalein in 50 mL ethanol and add 50 mL H_2O. (Expts 3, 9, 10, 13, 27, 29)　8.2–10　colorless to pink

Universal Indicator

Dissolve 0.003 g methyl orange, 0.015 g methyl red, 0.03 g bromothymol blue, and 0.036 g of phenolphthalein in 100 mL of 60% ethanol. Universal indicator solution can also be purchased from Fisher Scientific. (Expts 12, 13, 24)

Appendix **C**
Pure Substances

Elements	Experiment(s)	Organic Chemicals	Experiment(s)
Aluminum	1, 10, 14, 19, 21	Acetaldehyde	40
Bromine	40	Acetic anhydride	29
Carbon (graphite)	32	Acetone	2, 4, 17, 32, 38, 39A, 40
Chromium	21	Adipic acid	28
Copper wool	5	Ascorbic acid	39C
Copper strip	1, 13, 14, 21, 22, 31, 32	*iso*-Amyl alcohol	40
Iron wire	1, 14, 21, 31	Anthracene	20
Lead wire/strip	1, 21, 22, 31	Benzophenone	20
Magnesium ribbon	5, 7, 10, 13, 14, 22, 31	Biphenyl	20
Manganese	21	*t*-Butanol	20
Nickel wire/strip	1, 14, 21, 31	Camphor	20
Silicon	1, 10	Citric acid	28
Silver wire/strip	31	Cooking oil	29
Sodium	10	Cyclohexane	2, 17, 20
Sulfur	10, 20	*p*-Dibromobenzene	20
Tin strip	1, 21, 31	*p*-Dichlorobenzene	20
Zinc strip	1, 13, 14, 21, 22, 31	Diphenylamine	20
		Ethanol, 95%	1, 2, 28, 29, 37, 38, 39A, 40
		Ethanol, absolute	17
		Ethylenediamine	38
		Fumaric acid	28
		Hexane	17, 40
		1-Hexyne	40
		Maleic acid	28
		Malonic acid	28
		d,l-Mandelic acid	28
		Methanol	1, 17, 40
		Methylamine	40
		Mineral oil	10
		Naphthalene	20
		p-Nitrotoluene	20
		Oxalic acid	28
		1-Pentene	17, 40
		2-Pentene	40
		Phthalic acid	28
		1-Propanol	1
		iso-Propanol	17
		Salicylic acid	29, 40
		Succinic acid	28
		Sucrose (sugar)	13
		Tartaric acid	28
		Toluene	1, 20, 33, 40
		Xylene	40

Inorganic Chemicals (Knowns and Unknowns)

Compounds	Experiment(s)
Acetic acid, glacial, 17.4 M	40
Aluminum sulfate•18H$_2$O	21
Ammonia, conc, 14.8 M	4, 24, 34, 35, 38
Ammonium chloride	12, 13, 21
Ammonium nitrate	21
Ammonium sulfate	21, 37
Barium chloride•2H$_2$O	6, 8, 11, 12
Calcium carbonate	7, 12, 18, 26
Calcium chloride•2H$_2$O	11
Calcium hydroxide	13, 21, 34
Calcium oxide	13, 34
Calcium sulfate•2H$_2$O	6
Cobalt(II) chloride•6H$_2$O	12
Copper(II) bromide	24
Copper(II) chloride•2H$_2$O	11
Copper(II) sulfate•5H$_2$O	6, 12, 38
Hydrochloric acid, conc, 12.1 M	
	11, 13, 24, 36, 38
Iron(II) ammonium sulfate•6H$_2$O	39A, 39C
Iron(III) chloride•6H$_2$O	12
Iron(II) sulfate•7H$_2$O	37
Lithium chloride	11, 21
Lithium hydroxide	21
Magnesium oxide	26
Magnesium sulfate•7H$_2$O	6
Manganese(IV) oxide	22
Nickel(II) chloride•6H$_2$O	12, 38
Nitric acid, conc, 15.7 M	5, 14, 35, 37
Phosphoric acid, 85%	13
Potassium bromide	10, 21, 24
Potassium chloride	11, 21
Potassium dichromate	37
Potassium hydrogen phthalate (primary standard)	9, 28
Potassium hydrogen tartrate	28
Potassium hydroxide	21
Potassium iodate (primary standard)	15
Potassium iodide	10, 15, 21, 32
Potassium permanganate	16
Potassium nitrate	21
Potassium nitrite	14
Potassium sulfate	21
Silver nitrate	12

Compounds	Experiment(s)
Sodium bismuthate, 80% purity	35
Sodium bromide	21, 32
Sodium carbonate	12, 21
Sodium carbonate•H$_2$O	6
Sodium chloride	10, 11, 18, 21, 32
Sodium fluoride	10
Sodium hydrogen carbonate	18, 26
Sodium hydrogen oxalate	28
Sodium hydrogen sulfate	28
Sodium hydroxide	9, 21
Sodium iodide	13, 21
Sodium nitrate	21
Sodium oxalate	16
Sodium phosphate•12H$_2$O	8, 12
Sodium sulfate	21
Sodium sulfite	10
Sodium thiosulfate•5H$_2$O	15, 21
Strontium chloride•6H$_2$O	11
Sulfamic acid	28
Sulfuric acid, conc, 17.8 M	10, 13, 29, 33, 40
Zinc sulfate•7H$_2$O	6

Commercial Chemicals	Experiment(s)
Aluminum (can, foil, ...)	19, 37, 39A
Alka-Seltzer	18
Ammonia, household	13
Antacids, assortment	26
Aspirin	29
Bleach, household	10, 14, 15
Bleach, powder or mildew remover	15
Cooking oil	2, 29
Cotton	40
Detergent (any)	13
Gasoline	40
HTH disinfectant	15
Lemon juice	13
Mildew remover	15
Mineral oil	10
Styrofoam balls	DL3
Vinegar, assortment	13, 27
409™	13
7-UP™	13

Special Equipment

Equipment	Experiment(s)	Equipment	Experiment(s)
Aluminum foil	10, 16, 17, 39A, 39C	Paper clips or staples	4
Ammeter (0.1–1 A)	32	Pipet, dropping or Beral	5, 23, 33
*Balances, ±0.01 g		Pipet and pipet bulb, 1 mL	23, 25, 31
*Balances, ±0.001 g		Pipet and pipet bulb, 2 mL	2, 22, 39C
Battery (dc power supply), 9 V transistor		Pipet and pipet bulb, 5 mL	1, 2, 22, 23, 25
with connecting wires	32	Pipet and pipet bulb, 10 mL	
Barometer	17, 18, 19		15, 16, 23, 25, 26, 39C
Beaker, 15–20-mL	23	Pipet and pipet bulb, 25 mL	15, 25, 26, 28, 30
Beaker, 600-mL	4, 17, 20	pH meter and electrodes	28
Beaker, 800-mL	13, 19	Plastic wrap	4
Beaker, 1-L	4, 19	Pneumatic trough	18, 19
Boiling chips	2, 17, 21	Polystyrene coffee cups	21
Büchner funnel, filter flask, and flask		Resistor, variable (optional)	32
clamp	29. 37, 38, 39A	*Ring stands	
Buret, 50-mL, buret brush, and clamp		Reagent bottle, polyethylene, 500-mL	9
	9, 15, 16, 23, 26, 27, 28, 29, 30, 39B	Rubber band	2, 17, 37
Buret, gas (optional), 50-mL	19	Rubber policeman	8
*Bunsen burner with tubing		Safety shield	10
Centrifuge	5, 33, 34, 35	Spectroscope (optional)	11
Chromatography paper	4	Spectrophotometer (visible) with cuvets	
Clay triangle	6, 7		25, 39C
*Clamps, assortment		*Stoppers, rubber, assortment	
Cobalt glass plate (for K test)	11, 34	Steel wool	5, 7, 10, 13, 14, 22, 31, 32
Crucible and lid and tongs	6, 7	Test tube, 200-mm	
Desiccator (or desicooler)	6, 7, 9, 15		10, 13, 18, 19, 20, 21, 25, 39A
Drying oven	8, 9, 15, 17, 28	*Test tube clamps	
Erlenmeyer flask, 250-mL	9, 26, 27, 29, 30	*Thermometer, –10°C to 110°C	
Erlenmeyer flask, 125-mL		Thermometer, 360°C	2, 29
	16, 26, 27, 29, 30	Timer	20, 22, 23
Evaporating dish	13, 34, 35, 36	*Tubing, rubber, assortment	
*Filter paper, assortment		*Tubing, glass, assortment	
Filter paper, fine porosity	8	U-tube	19, 32
Flame test wire	11, 34, 36	Voltmeter, 0-3 V with connecting wires	31
Fume hood	5, 29, 37	Voltmeter, expanded scale (optional)	31
Gas collecting bottles	13	Volumetric flask, 100-mL	15, 16, 25, 31, 39C
*Gas delivery tubes, assortment.		Volumetric flask, 50-mL	39C
See Appendix A of the laboratory manual		Volumetric flask, 10-mL	25
Glass plate	13, 18, 19	Watchglass	8, 11, 12, 29, 34, 37, 38
Hot plate	29, 37	Water aspirator, attached to cold water outlet	
Ice	20, 22, 29, 37, 38	Weighing paper	
*Iron support rings			8, 9, 15, 16, 18, 20, 21, 28, 29, 39A, 39B, 39C
Kimwipes	25, 28, 39C	6-Well plate and Beral pipets	23
Leveling tank or 4-L Beaker	18, 19	24-Well plate and Beral pipets	
*Litmus paper, red/blue			3, 10, 12, 13, 14, 22, 24, 31, 38
Melting point (capillary) tubes	2, 4, 29, 37	Wire stirrer	20, 21
Meter ruler	4	Universal indicator chart	12, 13, 24
Molecular models	DL3	Wood splints	13
Mortar and pestle	15		

---

* Items used in ten or more experiments.